Congratulations

You are the lucky winner of this valuable autographed First Edition.

I hope it will give you endless hours of reading pleasure.

Chris Darke

Calcium Metabolism:
Comparative Endocrinology

Calcium Metabolism: Comparative Endocrinology

Edited by
J Danks, C Dacke, G Flik and C Gay

Published by
BioScientifica

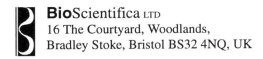 **Bio**Scientifica LTD
16 The Courtyard, Woodlands,
Bradley Stoke, Bristol BS32 4NQ, UK

©1999 BioScientifica Ltd

All rights reserved. No part of this book may be reproduced in any form or by any means without written permission from the publishers.

British Library Cataloguing in Publication Data
A CIP catalogue record for this book is available from the British Library.

ISBN 1 901978 05 2

Cover design by Argent

Printed in Great Britain by Cromwell Press

The IInd International Satellite Symposium on Calcium Metabolism: Comparative Endocrinology

held in conjunction with the
Second Joint Meeting of the American Society for Bone Mineral Research
and the International Bone and Mineral Society

30 November 1998, San Francisco, CA, USA

Preface

This book contains collected papers from the International Satellite Symposium on the Comparative Endocrinology of Calcium Regulation, the second satellite symposium to be held in recent years in conjunction with the main tri-annual conference of the International Bone and Mineral Society (IBMS), a joint meeting this time with the American Society for Bone and Mineral Research (ASBMR).

The previous satellite meeting took place in Melbourne in 1995 and the proceedings were published as a book the following year by The Journal of Endocrinology Ltd and entitled *The Comparative Endocrinology of Calcium Regulation*. The latest satellite was organised by Janine Danks from Melbourne, Carol Gay from Pennsylvania, Gert Flik from Nijmegen and Chris Dacke from Portsmouth, and held in November 1998 in the impressive venue of the California Academy of Sciences in Golden Gate Park, San Francisco. Delegates attended from North America, Europe, Japan and Australia. There were 27 papers in the form of overview talks, and oral and poster presentations.

At this symposium, the Invertebrate Overview paper was given by Gert Flik from Nijmegen and covered mainly crustacean calcium metabolism. This was followed by two interesting presentations on the mechanisms of calcium transport and moulting cycles and also genetic aspects of carapace formation in crustacea. Thrandur Bjornsson provided the fish overview and summarised much of the recent work in salmonid species emanating from his group in Goteberg. This was followed by a series of essentially fish papers on osteocalcin and response to stress, the Ca^{2+} receptor in flounder tissues and parathyroid hormone related peptide (PTHrP) receptors in cartilaginous and teleost fish. Due to illness the reptile/avian overview speaker was unable to attend and Chris Dacke stepped into the breach and gave a talk about avian medullary bone, although the original programmed talk is published here. A good range of papers on the effects of PTH and calcitonin in iguanas, dietary calcium/phosphorus ratios and bone quality in broiler chickens and the role of nitric oxide as a mediator of avian osteoclast activity and comparative studies of vitamin D-binding proteins in indigenous Australian species of birds and reptiles followed. In the latter part of the afternoon Tom Rosol from Ohio, enlivened us with a fascinating and highly comparative overview talk on aspects of mammalian calcium and bone regulation. This was followed by presentations on the role of PTHrP on regeneration of stag antlers, dietary phosphate deprivation in Romanian pigs, fracture repair in rats, foetal-placental calcium regulation in mice and, finally, the strange calcium metabolism of rabbits which happily run their plasma calcium levels about half as high again as most other vertebrate species.

By 5.30 p.m. the delegates were showing signs of exhaustion, having been on the go since 8.00 a.m. that day, many of them still suffering from jetlag. So, while they looked over the posters, they sampled local wines and the excellent San Francisco steam beer specially selected by the organisers. At 6.30 p.m. we bussed the by now relaxed crowd of delegates to a Chinese restaurant where they enjoyed a banquet of authentic delicacies washed down with more Californian wine. After dinner, we were treated to a thought-provoking summary of the day's scientific events by Howard Bern, the eminent comparative endocrinologist from Berkeley, who had attended the whole meeting. We are deeply grateful to Howard who, as always, rose to the occasion in his own inimitable style.

At the end of the symposium, several of the delegates commented on how much they had enjoyed the informative nature and informal atmosphere of the meeting. We are already looking forward to the next of these meetings which will be held in Faro, Portugal as a satellite of the XIVth IBMS meeting in Madrid in 2001.

The organisers are very grateful to Eli Lilly (USA), Procter and Gamble (USA), ASBMR, IBMS, Glaxo Wellcome (USA), Merck Sharpe and Dohme (Australia) and Schering (UK) who all provided generous support and sponsorship for our meeting.

Janine Danks, Chris Dacke, Gert Flik and Carol Gay

Contents

Part 1   Invertebrates

1   Calcium regulation in invertebrates
    G Flik, C Haond and C Lucu — 3

2   Physiological and molecular characterization of the
    calcium pump: evolutionary considerations
    M G Wheatly and Z Zhang — 13

3   Formation of the calcified exoskeleton in the prawn,
    Penaeus japonicus
    P Persson, T Ikeya and T Watanabe — 21

Part 2   Fish

4   Calcium balance in teleost fish: transport and endocrine
    control mechanisms
    *B Th Björnsson, P Persson, D Larsson, S H Jóhannsson
    and K Sundell* — 29

5   Osteocalcin response to environmental stressors
    *P E Patterson-Buckendahl, Z M G S Jahangir, M Rusnak
    and R Kvetnansky* — 39

6   The calcium-sensing receptor in fishes
    *P M Ingleton, P A Hubbard, J A Danks, G Elgar,
    R A Sandford and R J Balment* — 45

7   Parathyroid hormone-related protein: localisation in
    cartilaginous fish tissues
    *M K Trivett, T I Walker, J G Clement, P M Ingleton,
    T J Martin and J A Danks* — 49

8   Molecular cloning of cDNAs encoding three distinct
    receptors for parathyroid hormone (PTH)/PTH-related
    peptide in the zebrafish
    D A Rubin and H Jüppner — 59

Part 3   Reptiles and birds

9   Regulation of the calcium-sensing and parathyroid
    hormone receptor genes in the chick
    *M Pines, N Yarden, S Ben-Bassat, I Lavelin and
    R M Leach* — 67

10	Effects of daily administration of human parathyroid hormone (1-34) or salmon calcitonin in green iguanas (*Iguana iguana*) T J Rosol, J L Taylor, D G Fischbach, V Matkovic, M D Eberts, S N Huff and K M Morgan	75
11	Influence of dietary calcium and phosphorus on skeletal quality of the modern broiler chicken B Williams, D Waddington and C Farquharson	81
12	Measurement of nitric oxide production from isolated single chick osteoclasts using a porphyrinic microsensor S F Silverton, O A Adebanjo, B S Moonga, G D Markham, T Malinski, J V Johnson and M Zaidi	87
13	Variation in the concentration and binding affinities of the plasma vitamin D-binding protein C J Laing, G M Shea and D R Fraser	93
14	Japanese quail medullary bone, an *in vivo* model for bone turnover: effect of disodium pamidronate C G Dacke, J Sanz, K Foster, J Anderson and J Cook	99
15	Concentrations of 1,25-dihydroxyvitamin D3 and its plasma-binding protein are related to genetic susceptibility to tibial dyschondroplasia in the chicken M Lowe, C J Laing, W L Bryden and D R Fraser	103
16	Parathyroid hormone and estrogen effects on adhesion of chicken medullary bone osteoclasts T Sugiyama, S Kusuhara and C V Gay	107
17	Parathyroid gland hyperplasia in female snapping turtles (*Chelydra serpentina*) during egg-laying T J Rosol, P C Stromberg, J L Taylor, S W Fisher and J F Estenik	113

Part 4 Mammals

18	Mammalian calcium metabolism T J Rosol	119
19	Parathyroid hormone-related peptide may play a role in deer antler regeneration C Faucheux and J S Price	131

20	Pathophysiological effects of low dietary phosphate in pigs of southern Romania *J-L Riond, M Wanner, H Coste and G Pârvu*	139
21	Acceleration of rat femoral fracture healing by a synthetic thrombin peptide *D J Simmons, J Yang, S Yang, L X Bi, W L Buford, R T Turner, R Crowther and D H Carney*	145
22	Regulation of fetal-placental calcium metabolism *C S Kovacs, B Lanske, N R Manley and H M Kronenberg*	153
23	Endocrinology of calcium metabolism in rabbits *R Brommage*	159
24	Markers of bone resorption and formation during lactation in dairy cows *A Liesegang, M-L Sassi, J Risteli, M Kraenzlin, M Wanner and J-L Riond*	165
25	Cyclosporin A induces dentine resorption via a nitric oxide-cGMP pathway in osteoblast/osteoclast co-cultures: mechanistic insights *V S Shankar, S Yang, J Yang, M T Chapa, D J Simmons and S Wimalawansa*	169
26	Placental vitamin D regulates maternal calcium metabolism during pregnancy *H Fukuoka, M Haruna, C S Kim, M Fujita and Y Maekawa*	173
27	A mid-region parathyroid hormone-related peptide (PTHrP) stimulates the absorption of calcium ions from the ovine rumen *D R Wadhwa, A D Care and A F Stewart*	179

Part 5 Symposium Summary and Overview

28	Symposium summary and overview *H A Bern*	185

Author index   189

Subject index   191

Part One

Invertebrates

Calcium regulation in invertebrates

G Flik[1], C Haond[2] and C Lucu[3]

[1]Department of Animal Physiology, Faculty of Science, University of Nijmegen, Toernooiveld 1, 6525 ED Nijmegen, The Netherlands, and [2]Laboratoire d'Ecophysiologie des Invertébrés, Université Montpellier II, Montpellier, France and [3]Institute Ruder Boskovic, Rovinj, Croatia

Introduction

Studies on invertebrates have contributed significantly to our understanding of calcium regulation and have provided us with stimulating models in cell biology, biomineralization, endocrinology, physiology and membrane biology (e.g. Rasgado-Flores & Blaustein 1987, Falini et al. 1996, Coast & Webster 1998, Dipolo & Beaugé 1998). In this paper we will focus on regulated calcium transport in a particularly interesting group, the crustacea. The topic of calcification of a crustacean exoskeleton and its consequences for acid–base regulation and $CaCO_3$ stores will be addressed by Persson in this volume. Crustacea have a phenomenal capacity to handle calcium, a requirement for their abrupt growth that proceeds via moulting cycles. After each ecdysis, the protective carapace, which may form the basis for their evolutionary success, needs to be impregnated with ever increasing loads of $CaCO_3$ (calcite). To enlarge the whole body calcium pool for growth, the animal must realize a huge net influx of Ca^{2+} as pressure for remineralization is high because of the imminent threat of predation. To this end, Ca^{2+} is absorbed at high rates from the environment (Lucu 1994) either directly by epithelia in the branchial cavity or indirectly via the intestinal epithelium from ingested water or food or from stored Ca^{2+} deposits. Crustaceans also economize by recycling part of their carapace using Ca^{2+} acquired during growth. Ca^{2+} is reabsorbed via the cuticular hypodermis epithelium during premoult and is temporarily deposited in association with the gut; interestingly, in the postmoult stage the hypodermis mineralizes the cuticle and thus the net Ca^{2+} flow in this epithelium can be reversed (the epithelium becomes secretory). In many instances, shed exuviae, which may still contain a considerable amount of the original minerals (30–70%), are eaten to optimize recycling. Significant parts of these phenomenal flows of Ca^{2+} proceed transcellularly and at the same time cellular Ca^{2+} homoeostasis must be guaranteed. For an elaborate review of Ca^{2+} homoeostasis in crustacea we refer to recent reviews by Wheatly and coworkers (Wheatly 1989, 1996) and the paper in this volume by Wheatly. Even in the intermoult phase (Cameron 1989), i.e. the relatively 'quiet' period between moults, crustacea still transport Ca^{2+} at high rates to counteract diffusive losses. This chapter will review and rely mainly on studies of intermoult crustaceans as reports on pre- and post-moult animals are relatively scarce.

Aquatic species have access to an infinite source of dissolved calcium. As ambient water Ca^{2+} concentrations exceed intracellular Ca^{2+} concentrations (typically less than 1 µM free Ca^{2+}) and the cell interior is negatively charged, the animal in principle can utilize the inwardly directed electrochemical gradient for ambient Ca^{2+} by regulating entry of Ca^{2+} into its Ca^{2+}-transporting epithelia associated with the branchial cavity. A situation comparable with that in fish seems to exist here: at present, consensus exists that Ca^{2+} uptake in fish is under a predominantly inhibitory control over Ca^{2+} entry from ambient water via the mitochondria-rich cells in the branchial epithelium. The hormone stanniocalcin exerts antihypercalcaemic actions on the gills by limiting Ca^{2+} entry via second-messenger-operated Ca^{2+} channels in the apical membrane of the mitochondria-rich cells (Flik & Verbost 1996, Verbost et al. 1996). In fish, where extirpation of the glands that produce stanniocalcin is feasible, the surgical procedure is followed by a rapid rise in plasma Ca^{2+}, the rise resulting from increased and uncontrolled inflow of Ca^{2+} from ambient water. Accordingly, in seawater (as compared with fresh water) the turnover of the hormone is higher and the target volume for stanniocalcin increases, indicating that the high calcium activity in seawater (and the overall electrochemical conditions for Ca^{2+} transport) require enhanced control over Ca^{2+} entry (Hanssen et al. 1992). Fish control plasma Ca^{2+} almost perfectly and independently of ambient Ca^{2+} levels, and calculated transepithelial potentials deviate from equilibrium potentials such that the Ca^{2+} influx observed in fresh water as well as seawater conditions must be based on active energized transport of Ca^{2+} (Perry & Flik 1988, Flik & Verbost 1996,). On the other hand, the passive driving force for Ca^{2+} flow in fish is directed outward in both fresh water and seawater. In agreement with this model, it was shown that the branchial epithelium of fish harbours Ca^{2+}-ATPase and Na^+/Ca^{2+}-exchanger molecules that guarantee energized extrusion of Ca^{2+} through the mitochondria-rich cells as the basis for branchial Ca^{2+} transport (Flik & Verbost 1996).

For many years it was thought that the hormone stanniocalcin was unique to fish, as its actions were convincingly demonstrated only in fish. However, stanniocalcin immuno-crossreactivity has been demonstrated in molluscs and higher vertebrates (Wendelaar Bonga et al. 1989), and recently the human stanniocalcin gene was identified (Chang et al. 1998). These studies indicate that the hormone is a phylogenetically well-conserved signal molecule. An important concept is that stanniocalcin is no longer considered to be strictly a Ca^{2+}-regulating hormone, as demonstrated by the fish gill studies; it is now also known to have phosphate-regulatory effects in fish and mammalian kidney cells (Lu et al. 1994). Whether crustacean species express a gene for a stanniocalcin molecule and utilize such a molecule in Ca^{2+} regulation remains to be investigated. Little is known about phosphate regulation in crustacea (Glynn 1968) and, as will be discussed below, the mechanisms for apical membrane uptake of Ca^{2+} may be basically different from those seen in fish. Also, crustacean species hyperregulate haemolymph Ca^{2+} levels, but studies on this hyperregulation have mainly addressed total Ca^{2+} levels in the haemolymph rather than the electrophysiologically important Ca^{2+} activity.

Ca^{2+} absorption via crustacean gills

On initial confrontation of the (ultra-)structure of crustacean gills or associated structures, the (non-calcified) cuticle appears as a solid barrier. However, this protective structure functions as an exoskeleton that is fully permeable to Ca^{2+} and Na^+ ions (Neufeld & Cameron 1993) and therefore does not form a barrier to the passage of Ca^{2+} ions from medium into transporting cells.

For intermoult blue crab, a value of 10 μmol/h per kg was reported, for animals in both seawater and diluted seawater where active Ca^{2+} transport is a requirement (Neufeld & Cameron 1993, 1994); these values are comparable with Ca^{2+} fluxes in fish gills, specifically, 14 μmol/h per kg for tilapia (Flik *et al.* 1985) and 14.65 μmol/h per kg for trout (Perry & Flik 1988). In postmoult gills, however, an inward Ca^{2+} flux of over 500 μmol/h per kg has been reported for shore crab gills, indicating the adaptive capacity of the organs (Greenaway 1983).

As shown in Table 1, haemolymph total calcium levels always surpass ambient calcium levels, but Ca^{2+} levels are hyperregulated only in media containing 5 mmol/l (about half the Ca^{2+} concentration found in seawater) or less. The calculated equilibrium potentials in brackish or fresh water environments appear to be significantly lower than the transbranchial potentials under those conditions, and therefore energized transport of Ca^{2+} is required for hyperregulation under those conditions. In isolated and perfused gills (under symmetrical conditions, i.e. perfused with the saline of the bath) of brackish-water-adapted shore crab *Carcinus maenas*, the transbranchial potential instantaneously resets to about 0 mV when calcium is selectively omitted from the perfusate; removing calcium from the bathing medium has no substantial effect on the transbranchial potential (Lucu 1994, C Lucu, unpublished). Clearly, the regulation of the haemolymph Ca^{2+} concentration is crucial in determining the conditions for transport of Ca^{2+} across the gills of crabs, a situation strikingly different from fish where the transepithelial potential is determined by ambient calcium levels (McWilliams 1982).

The apical membrane compartment

Ca^{2+} absorption across the crustacean gill as a function of ambient calcium levels is described by a saturation curve, indicating carrier-mediated transport. Typically, Ca^{2+} inflow in *C. maenas* and *Callinectes sapidus* kept in brackish water ($[Ca^{2+}]=5$ mmol/) shows half-maximal saturation at an external Ca^{2+} concentration of around 10 mmol/l (Greenaway 1983). For the freshwater crayfish, *Austropotamobius*, and the gammarid, *Gammarus pulex*, these values are 0.1 and 0.3 mmol/l respectively (Greenaway 1974), indicating that the apical membrane Ca^{2+} carrier process operates at 50% capacity at prevailing Ca^{2+} concentrations. The carriers in the basolateral membrane compartment have affinties for Ca^{2+} in the (sub-)micromolar range and thus do not determine this transport process. The kinetics of Ca^{2+} uptake from the water allow for a direct regulatory control by small variations in ambient calcium on Ca^{2+} uptake. It should be mentioned that in very soft calcium-poor waters (0.075 to 0.1 mM), crustacean growth

Table 1. Haemolymph calcium and electrochemical conditions for branchial Ca^{2+} transport in crustacea.

Species	$[Ca]_M$ (mM)	$[Ca]_H$ (mM)	$[Ca^{2+}]_H$ (mM)	TBP (mV)	P_{EQ} (mV)	Reference
Pacifastacus leniusculus	0.05	19	9.5*	-13.9	-66.2	Wheatly (1989)
	6.82	15	7.5*	0.5	-1.2	
	15.6	22.5	11.3*	1.0	4.1	
A. pallipes	0.6	12.2	6.3	-11.4	-29.7	Greenaway (1974)
C. sapidus	1.6 to 2.1	11.8	7.9	-12.1	-16.7 to -20.1	Neufeld &
	11.0	12	8.2	-2.2	3.7	Cameron (1993,1994)
C. maenas	3.6	9.2	5.1	-0.8 to 2.4	-3.4	Lucu & Flik (1999) Wheatly (1999)†
	5.2	11.3	5.1	n.d.	0.12	
	7.1	11.3	5.6	n.d.	2.89	
	10.7	12.7	6.8	-0.9	5.2	

Total ($[Ca]_H$) and ionic ($[Ca^{2+}]_H$) calcium concentrations in in haemolymph of crustacea acclimated to media with various calcium concentrations ($[Ca]_M$). Measured transbranchial potentials (TBP, in mV) are compared with calculated Nernst equilibrium potentials (P_{EQ}= 29.07× $[Ca]_M$ $[Ca^{2+}]_H$ mV).
*If data on haemolymph Ca^{2+} concentrations were not available from the literature, values were calculated assuming 50% of the total calcium being in ionised form.
†this volume

is inhibited. In such conditions the ambient sources of calcium indeed become limiting and the animals must rely on calcium ingested and stored in the body (Huner et al. 1976).

Studies on brush border membrane vesicles isolated from crustacean gills have revealed the presence of a unique electrogenic $2Na^+/H^+$ exchanger (Shetlar & Towle 1989, Towle et al. 1998). This carrier is of widespread occurrence as it was also demonstrated in hepatopancreas of the fresh water and marine crustacea (Ahearn & Clay 1989, Ahearn et al. 1994), in crustacean antennal glands (Ahearn & Franco 1990) and in echinoderm gastrointestinal epithelium (Ahearn & Franco 1991). In membrane vesicles, Na^+ uptake is markedly enhanced by an outwardly directed proton flux and the negative membrane potential (negative sign inside relative to the outside of the right-side-out oriented vesicles) resulting from the activity of this carrier process indicates that this countertransport is electrogenic. The Na^+ dependence of this transport process reveals cooperativity for Na^+ ions (Na^+ transport is described by a sigmoidal curve), indicating at least two Na$^+$ sites; moreover, this transport is sensitive to extravesicular amiloride in line with Na$^+$ sites in the extracellular domain. Finally, one intravesicular H$^+$-binding site was demonstrated that also binds cytosolic Na$^+$. Interestingly, in an antennal gland preparation, Ca^{2+} competitively inhibits electrogenic 2Na$^+$/H$^+$-exchange activity (Ahearn & Franco 1990). In the lobster antennal gland, Ca^{2+} was demonstrated to interact with the 2Na$^+$/H$^+$-exchanger on its low- and high-affinity binding sites for Na$^+$, i.e. the same sites where amiloride interacts with this protein. The same carrier protein may thus mediate four types of countertransport: 2Na$^+$/H$^+$, 2Na$^+$/Na$^+$, Ca^{2+}/H$^+$ and Ca^{2+}/Na$^+$. Importantly, an additional amiloride-insensitive electroneutral Ca^{2+}/2Na$^+$ exchange was detected. The capacity of the amiloride-sensitive Ca^{2+} flux (V_{max}=147.89 nmol/min per mg protein) is roughly three times higher than that of the amiloride-insensitive Ca^{2+} flux (Ahearn & Franco 1993), signifying the imporance of the electrogenic transporter. Being electrogenic, this carrier is voltage-dependent and may be coupled indirectly to the pivotal Na$^+$/K$^+$-ATPase of the cell that governs its electrical behaviour. Finally Ca^{2+} ions may move into the cell via verapamil-inhibitable Ca^{2+} channels: in antennal gland brush border membranes verapamil inhibits part of the Ca^{2+} influx stimulated by outwardly directed H$^+$ and Na$^+$ gradients. This indicates the presence of verapamil-sensitive Ca^{2+} channels (Ahearn & Franco 1993).

Using a monoclonal antibody raised against antigen in American lobster hepatopancreas brush border membrane, it was shown by a combination of Western blot analysis and vesicle transport studies, in which the antibody was shown to inhibit 2Na$^+$/H$^+$ exchange and Ca^{2+}/H$^+$ exchange, that the pertinent carrier protein in hepatopancreas, antennal gland, and gill epithelium involves a single protein species with a molecular mass of 185 kDa (Gert de Couet et al. 1993). An interesting hypothesis to test is whether the crustacean electrogenic 2Na$^+$ (Ca^{2+})/1H$^+$ (Na$^+$) exchanger works in concert with H$^+$-ATPase activity. For freshwater Chinese crab gills it was suggested that a V-type ATPase may provide a protonmotive force for Na$^+$ transport (Onken & Putzenlechner 1995). Import of Na$^+$ ions via a Na$^+$ channel coupled to a protonmotive force provided by an apical membrane H$^+$- ATPase activity is, at present, the most adhered to model for Na$^+$ transport in frog skin and fish gill (Harvey et al. 1991, Perry & Fryer 1997). The unique properties of the crustacean 2Na$^+$ (Ca^{2+})/1H$^+$ (Na$^+$) exchanger warrant testing this hypothesis, but its countertransport nature would at first glance make a direct dependence of Ca^{2+} transport on ATPase activity unlikely.

The basolateral plasma membrane compartment
Significant progress in demonstrating Ca^{2+}-transporting enzyme activity involved in Ca^{2+} extrusion from the cell was first made by increasing the percentage of inside-out vesicles of branchial basolateral plasma membrane preparations of posterior gills of the shore crab (Flik et al. 1994, Lucu & Flik 1998) and by applying the same procedure to epithelia of the branchial cavity of the lobster (Flik & Haond 1999). By using hypotonic shock treatment (Flik et al. 1994), sufficient resealed and inside-out oriented vesicles may be obtained to study Ca^{2+}-ATPase and Na$^+$/Ca^{2+}

exchanger-mediated Ca^{2+} uptake. These carriers have now been described for a variety of crustacean ion-transporting tissues (Wheatly 1996), but an intriguing question remains, which one of these carriers serves in a transporting epithelium?

To answer this question, one could first consider the kinetic parameters of the carriers (Flik *et al.* 1990). Ca^{2+}-ATPase is a high-affinity low-capacity carrier for Ca^{2+} and as a result cytosolic Ca^{2+} activity will be a major determining factor in the activity of this transporter. The exchanger on the other hand is a low-affinity high-capacity carrier for Ca^{2+}, and its activity is governed by ion conditions as well as membrane potential (Flik *et al.* 1990, Flik & Verbost 1996). Exact evaluation of the relative importance of the exchanger thus awaits physiological studies that evaluate ion activities and membrane potentials in Ca^{2+}-transporting cells. Indirect evidence for involvement of exchanger-based Ca^{2+} transport comes from studies showing that the postmoult calcification depends on secondary active (i.e. Na^+,K^+-ATPase-dependent) Ca^{2+} transport (Roer 1980, Towle *et al.* 1998). Biochemical assays on the other hand may not have the sensitivity required to demonstrate the involvement of exchange activity in Ca^{2+} transport. In a recent study (Lucu & Flik 1998), shore crab was challenged with dilute seawater conditions. When transferred from seawater and adapted to dilute seawater (10%; 3.6 mmol/l Ca^{2+}), the shore crab hyperregulates its haemolymph Ca^{2+} levels (5.1 mmol/l) by enhanced active uptake from the water. A 1.67-fold increase in total exchanger activity was calculated on the basis of increased protein synthesis in the gills. However, no adjustments of the exchanger density was observed in these animals (Ca^{2+} and Na^+ kinetics of the exchanger were unaffected by acclimation conditions). On the other hand, the ouabain-sensitive short-circuit current in a hemilamellar preparation of the posterior gills almost doubled and the Na^+,K^+-ATPase activity and α-subunit expression almost quadrupled. It was concluded that the large Ca^{2+}-transport capacity of the exchanger suffices for its operation in seawater as well as dilute seawater animals; as the Na^+ pump is upregulated so strongly in dilute seawater, the dependence of the exchanger on the Na^+ gradient created by Na^+,K^+-ATPase activity infers no risk for Ca^{2+} transport.

Given this situation, a solution for short-term regulation of Ca^{2+} uptake also seems warranted. Immediately after transfer to dilute seawater, haemolymph mineral content falls, but stabilizes after about 24 h. Interestingly, a rapid increase in Na^+,K^+-ATPase activity precedes this stabilization suggesting that the Na^+ pump readjusts haemolymph ion levels rapidly. Evidence was provided that this rapid increase in pump activity must be realized by activation of existing Na^+,K^+-ATPase molecules (α-subunit expression did not change) or enzyme recruitment (not all enzyme activity present in the gills is active as Na^+ pump). Indeed, the rapid increase in Na^+,K^+-ATPase activity could be correlated with decreasing cAMP levels in the same tissue, and thus activation of the enzyme requires downregulation of cAMP levels. These observations strongly point to a cAMP-dependent protein kinase A pathway in the regulation of the Na^+ pump in this tissue, as has been documented for a variety of vertebrate cells. Again, being coupled to the Na^+ pump, the exchanger may change its activity in parallel. Further investigations of regulation of Ca^{2+} transport in crustacean epithelia should consider the pivotal role of the Na^+ pump and its regulation in secondary active transport of Ca^{2+}.

Lobsters (*Homarus gammarus* and *Homarus americanus*) are representatives of crustaceans with trichobranchiate gills (shore and blue crabs have phyllobranchiate gills, with the posterior four pairs being specialized for ion transport). Considering the general pattern of decapod crustaceans, these gills are assumed to be specialised for gas exchange and waste excretion. In addition the branchial cavity shelters 7 pairs of well-developed lamellar epipodites which are internally lined with a monolayer of ion-transporting cells. Similar ion-transporting epithelium also covers the inner side of the branchiostegite within the branchial cavity (Haond et al. 1998). The response of these structures in the branchial cavity to exposure of the animal to dilute seawater (22%) were studied histologically and biochemically (Flik & Haond 1999). The lobster hyperregulates its haemolymph osmolarity and Ca^{2+} concentration when confronted with dilute seawater. The epithelia of the epipodites and branchiostegites undergo marked ultrastructural changes suggesting enhanced transport activity; specifically, the transport cells hypertrophy, increase their basal labyrinth volume and the height of their apical microvilli, and numerous vesicles appear in the cytoplasm. The thin epithelium of the gills on the other hand appear unchanged. These histological observations suggest that lobster gills apparently are not osmo/ ionoregulatory structures, whereas the epipodites and branchiostegites could be. On the other hand, biochemical analyses revealed that all three structures contain significant levels of Na^+,K^+-ATPase, Ca^{2+}-ATPase and Na^+/Ca^{2+}-exchange activities. Exposure to dilute seawater enhanced Na^+/Ca^{2+}-exchange and Ca^{2+}-ATPase activities in epipodites and branchiostegites; however, in the gills Ca^{2+} transporter activities decreased. The differential response of the gills (increased Na^+,K^+-ATPase activity, decreased Ca^{2+} transporter activity) and of the epipodites and branchiostegite suggests a specialization of the latter two structures for Ca^{2+} transport. The increased Na^+,K^+-ATPase activity in the gills may be explained by the enhanced need for waste excretion in conditions of enhanced metabolism as imposed by exposing the animal to dilute seawater. In future studies we will mount lobster epithelia of the epipodites and branchiostegites in Ussing chambers and investigate the role of the ouabain-sensitive Na^+ pump and the Ca^{2+} carriers in Ca^{2+} transport and in the adaptive response to dilute seawater.

Regulation of Ca^{2+} transport

With the demonstration of the pivotal role of Na^+,K^+-ATPase in Ca^{2+} transport in crustacean Ca^{2+}-transporting epithelia, a new area of regulation should be considered in Ca^{2+} transport such as (de-) phosphorylation of the Na^+ pump. Recent new insights from invertebrate models for Na^+/Ca^{2+}-exchange include inhibitory regulation by intracellular Ca^{2+} of a Na^+/Ca^{2+} exchanger 'Calx' in *Drosophila* (Hryshko et al. 1996), MgATP regulation of squid Na^+/Ca^{2+}-exchanger mediated by cytosolic proteins (DiPolo et al. 1997) and N-ω-phosphoarginine (i.e. metabolic) regulation of squid Na^+/Ca^{2+}-exchanger (Dipolo & Beaugé 1995), contributions that may have impact far ouside the field of invertebrate calcium regulation.

We are only at the brink of understanding endocrine control of calcium regulation in invertebrates, but important roles may be considered for calcitonin, calcitonin gene-related peptide, vitamin D metabolites and ecdysteron, hormones that become (more) active during the moulting cycle at times of calcium recruitment (Meyran et al. 1991, Wheatly 1996, Coast & Webster 1998). We hope that this work contributes to future comparative studies and that the tools (signals and targets) of invertebrate calcium regulation will interest those studying calcium regulation.

Acknowledgements

CL and CH received funding from Research School 'Milieu and Toxicologie' (M&T) Wageningen, The Netherlands to carry out research in the Nijmegen Laboratory.

References

Ahearn GA & Clay LP (1989) Kinetic analysis of electrogenic $2Na^+$-$1H^+$ antiport in crustacean hepatopancreas. *American Journal of Physiology* **257** R484-R493.

Ahearn GA & Franco P (1990) Na+ and calcium share the electrogenic $2Na^+$-$1H^+$ antiporter in crustacean antennal glands. *American Journal of Physiology* **259** F758-F767.

Ahearn GA & Franco P (1991) Electrogenic $2Na^+/H^+$ antiport in echinoderm gastrointestinal epithelium. *Journal of Experimental Biology* **158** 495-507.

Ahearn GA & Franco P (1993) Ca^{2+} transport pathways in brush-border membrane vesicles of crustacean antennal glands. *American Journal of Physiology* **264** R1206-R1213.

Ahearn GA, Zhuang Z, Duerr J & Pennington V (1994) Role of the invertebrate electrogenic $2Na^+/1H^+$ antiporter in monovalent and divalent cation transport. *Journal of Experimental Biology* **196** 319-335.

Cameron JN (1989) Post-moult calcification in the blue crab, *Callinectes sapidus*; timing and mechanisms. *Journal of Experimental Biology* **142** 285-304.

Chang AC, Jeffrey KJ, Tokutake Y, Shimamoto A, Neumann AA, & Dunham MA et al. (1998) Human stanniocalcin (STC): genomic structure, chromosomal localization, and the presence of CAG trinucleotide repeats. *Genomics* **47** 393-398.

Coast GM & Webster SG (1998) Recent advances in arthropod endocrinology. *Society for Experimental Biology Seminar Series* **65**.

Dipolo R & Beaugé L (1995) Phosphoarginine stimulation of Na^+/Ca^{2+} exchange in squid axons: a new pathway for metabolic regulation? *Journal of Physiology* **487** 57-66.

Dipolo R & Beaugé L (1998) Differential up-regulation of Na^+/Ca^{2+} exchange by phosphoarginine and ATP in dialysed squid axons. *Journal of Physiology* **507** 737-747.

DiPolo R, Berberian G, Delgado D, Rojas H & Beaugé L (1997) A novel 13 kDa cytoplasmic soluble protein is required for the nucleotide (MgATP) modulation of the Na/Ca exchange in squid nerve fibers. *FEBS Letters* **401** 6-10.

Falini G, Albeck S, Weiner S & Addadi L (1996) Control of aragonite or calcite polymorphism by mollusc shell macromolecules. *Science* **271** 67-69.

Flik G & Haond C (1999) Na,K-ATPase, Ca-ATPase and Na/Ca-exchange activities in gills, epipodites and branchiostegite of European lobster, *Homarus gammarus*: effects of exposure to dilute seawater. *Journal of Experimental Biology* In Press.

Flik G & Verbost PM (1996) Calcium homeostasis in fish. In *The Comparative Endocrinology of Calcium Regulation*, pp 43-53. Eds C Dacke, J Danks, I Caple & G Flik. Bristol: Journal of Endocrinology Ltd.

Flik G, Fenwick JC, Kolar Z, Mayer-Goston N & Wendelaar Bonga SE (1985) Whole body calcium flux rates in cichlid teleost fish *Oreochromis mossambicus* adapted to freshwater. *American Journal of Physiology* **249** R432-R437.

Flik G, Schoenmakers TJM, Groot JA & Van Os CH (1990) Calcium absorption by fish intestine: the involvement of ATP- and sodium-dependent calcium extrusion mechanisms. *Journal of Membrane Biology* **113** 13-22.

Flik G, Verbost PM, Atsma W & Lucu C (1994) Calcium transport in gill plasma membranes of the crab *Carcinus maenas*: evidence for carriers driven by ATP and a sodium gradient. *Journal of Experimental Biology* **195** 109-122.

Gert de Couet H, Busquets-Turner L, Gresham A & Ahearn GA (1993) Electrogenic $2Na^+/1H^+$ antiport in crustacean epithelium is inhibited by a monoclonal antibody. *American Journal of Physiology* **264** R804-R810.

Glynn JP (1968) Studies on the ionic, protein and phosphate changes associated with the moult cycle of *Homarus vulgaris*. *Comparative Biochemistry and Physiology* **26** 937-946.

Greenaway P (1974) Total body calcium and haemolymph calcium concentrations in the crayfish *Austropotamobius pallipes* (Lereboullet). *Journal of Experimental Biology* **61** 35-45.

Greenaway P (1983) Uptake of calcium at the postmoult stage by the marine crabs *Callinectes sapidus* and *Carcinus maneas*. *Comparative Biochemistry and Physiology* **75A** 181-184.

Hanssen RGJM, Mayer-Gostan N, Flik G & Wendelaar Bonga SE (1992) Influence of ambient calcium levels on stanniocalcin secretion in the European eel (*Anguilla anguilla*). *Journal of Experimental Biology* **162** 197- 208.

Haond C, Flik G & Charmantier G (1998) Confocal laser scanning and electron microscopical studies on osmoregulatory epithelia in the branchial cavity of the lobster *Homarus gammarus*. *Journal of Experimental Biology* **201** 1817-1833.

Harvey BJ, Lacoste I & Ehrenfeld J (1991) Common channels for water and protons at apical and basolateral cell membranes of frog skin and urinary bladder epithelia: effects of oxytocin, heavy metals, and inhibitors of proton-ATPase. *Journal of General Physiology* **97** 749-776.

Hryshko L V, Matsuoka S, Nicoll D A, Weiss J N , Schwarz E M, Benzer S & Philipson KD (1996) Anomalous regulation of the *Drosophila* Na^+/Ca^{2+} exchanger by Ca^{2+}. *Journal of General Physiology* **108** 67-74.

Huner JV, Kowalczuk JG & Avault JW (1976) Calcium and magnesium levels in the intermolt (C4) carapaces of three species of freshwater crawfish (Cambaridae:decapoda). *Comparative Biochemistry and Physiology* **55A** 183 - 185.

Lu M, Wagner GF & Renfro JL (1994) Stanniocalcin stimulates phosphate reabsorption by flounder renal proximal tubule in primary culture. *American Journal of Physiology* **267** R1356-R1362.

Lucu C (1994) Calcium transport across isolated gill epithelium of *Carcinus*. *Journal of Experimental Zoology* **268** 339-346.

Lucu C & Flik G (1999) Na,K-ATPase and Na/Ca-exchange activities in gills of hyperregulating *Carcinus maenas*. *American Journal of Physiology* **276** R490-R499

McWilliams PG (1982)The effects of calcium on sodium fluxes in the brown trout, *Salmo trutta*, in neutral and acid water. *Journal of Experimental Biology* **96** 439-442.

Meyran JC, Chapuy MC, Arnaud S, Sellem E & Graf F (1991) Variations of vitamin D-like reactivity in the crustacean *Orchestia cavimana* during the molt cycle. *General and Comparative Endocrinology* **84** 115-120.

Neufeld DS & Cameron JN (1993) Transepithelial movement of calcium in crustaceans. *Journal of Experimental Biology* **184** 1-16.

Neufeld DS & Cameron JN (1994) Effect of the external concentration of calcium on the postmoult uptake of calcium in blue crab (*Callinectes sapidus*). *Journal of Experimental Biology* **188** 1-9.

Onken H & Putzenlechner M (1995) A V-ATPase drives active, electrogenic and Na-independent Cl absorption across the gills of *Eriocheir sinensis*. *Journal of Experimental Biology* **198** 767-774.

Perry SF & Flik G (1988) Characterization of branchial transepithelial calcium fluxes in freshwater trout, *Salmo gairdneri*. *American Journal of Physiology* **254** R491-R498.

Perry SF & Fryer JN (1997) Proton pumps in the fish gill and kidney. *Fish Physiology and Biochemistry* **17** 363-369.

Rasgado-Flores H & Blaustein MP (1987) Na/Ca exchange in barnacle muscle cells has a stoichiometry of $3Na^+/1Ca^{2+}$. *American Journal of Physiology* **252** C494-C504.

Roer RD (1980) Mechanisms of resorption and desorption of calcium in the carapace of the crab *Carcinus maenas*. *Journal of Experimental Biology* **88** 205-218.

Shetlar RE & Towle DW (1989) Electrogenic Na+-proton exchange in membrane vesicles from crab (*Carcinus maenas*) gill. *American Journal of Physiology* **257** R924-R931.

Towle DW, Rushton ME, Heidysch D, Magnani JJ, Rose MJ, Amstutz A, *et al.* (1998) Sodium/proton antiporter in the euryhaline crab *Carcinus maenas*: molecular cloning, expression and tissue distribution. *Journal of Experimental Biology* **200** 1003-1013.

Verbost PM, Fenwick JC, Flik G & Wendelaar Bonga SE (1996) Cyclic adenosine monophosphate, a second messenger for stanniocalcin in tilapia gills? In *The Comparative Endocrinology of Calcium Regulation*, pp 55-69. Eds C Dacke, J Danks, I Caple & G Flik. Bristol: Journal of Endocrinology Ltd.

Wendelaar Bonga SE, Lafeber FPJG, Flik G, Kaneko T and Pang PKT (1989) Immunocytological demonstration of a novel system of neuroendocrine peptidergic neurones in the pond snail *Lymnaea stagnalis*, with antisera to the teleostean hormone hypocalcin and mammalian parathyroid hormone. *General and Comparative Endocrinology* **75** 29-38.

Wheatly MG (1989) Physiological responses of the crayfish *Pacifastacus leniusculus* (Dana) to environmental hyperoxia. I. Extracellular acid-base and electrolyte status and transbranchial exchange. *Journal of Experimental Biology* **143** 33-51.

Wheatly MG (1996) An overview of calcium balance in crustaceans. *Physiological Zoology* **69** 351-382.

Wheatly MG (1998) Calcium homeostasis in crustacea: the evolving role of branchial, renal, digestive and hypodermal epithelia. *Journal of Experimental Zoology* (In Press).

Zanders IP (1980) Regulation of blood in *Carcinus maenas* L. *Comparative Biochemistry and Physiology* **65A**, 97-108.

Physiological and molecular characterization of the calcium pump: evolutionary considerations

M G Wheatly and Z Zhang

Department of Biological Sciences, Wright State University, Dayton, Ohio 45435, USA

Introduction

In recent years our laboratory has focused on developing the molting cycle of the fresh water crayfish as a model to study cellular mechanisms of calcium translocation and their molecular regulation (Wheatly 1999). Central to calcium homeostasis in all cells is the calcium pump (calcium adenosine triphosphatase, Ca^{2+}-ATPase) which transports calcium against its electrochemical gradient using free energy from hydrolysis of ATP. The calcium pump is found at two locations within the cell, either on external membranes (plasma membrane calcium ATPase, PMCA) where it effects Ca^{2+} extrusion from the cell (either to fine tune intracellular $[Ca^{2+}]_i$ or for efflux) or on internal membranes (sarco/endoplasmic reticulum calcium ATPase, SERCA) where it effects $[Ca^{2+}]_i$ sequestration. Extending characterization of this ancient gene/protein to invertebrates enables us to explore the molecule in an evolutionary context.

Physiological characterization of the calcium pump

PMCA function has been most often studied using basolateral membrane vesicles of epithelia which transport Ca^{2+} in the basolateral direction (gut, kidney, gill). It has also been identified on apical membranes in postmolt crustacean hypodermis (Wheatly 1999). PMCA activity has been measured as ATP-dependent $^{45}Ca^{2+}$ uptake into tightly sealed inside out basolateral membrane vesicles (IOVs, Table 1) in epithelia from species inhabiting a range of calcium environments including intermolt crustacean gill, hepatopancreas and antennal gland, fish gill and intestine, and mammalian kidney and gut. Flow cytometry is emerging as a new technology for measuring Ca^{2+} uptake into IOVs via PMCA as increased fluorescence of the Ca^{2+} indicator fluo-3. Preliminary studies reveal comparable kinetics and pharmacology (Telford & Miller 1996, J R Weil, W Telford & M G Wheatly, unpublished results).

The ATP-dependent $^{45}Ca^{2+}$ uptake mechanism in IOVs displays the following common characteristics: homogeneity; high affinity for calcium with half maximal saturation at $[Ca^{2+}]_i$; ATP dependence with $K_{0.5}$ of 0.01-2 mM; prevention by the Ca^{2+} ionophore A23187; inhibition (80-100%) by sodium orthovanadate, a non-specific 'P'-type ATPase inhibitor; temperature sensitivity; calmodulin dependence; and Na^+ dependence. In contrast, uptake is unaffected by the Na/K-ATPase inhibitor ouabain,

Table 1 ATP-dependent Ca^{2+} uptake into basolateral membrane vesicles of transporting epithelia of different species: kinetic parameters and maximal Ca^{2+}-transport capacity

Species	Ref	Medium	Tissue	Temp (°C)	K_m (μM)	J_{max} (nmol/min per mg protein)	IOVs (%)	Transport capacity (nmol/min per mg protein)
Invertebrates								
Shore crab	1	50%SW	Gill	37	0.150	1.73	22	7.87
Lobster	2	SW	Hepatopancreas	20	0.065	8.03	6	132.97
Crayfish	3	FW	Gill	37	0.280	0.11	17	0.63
			Hepatopancreas		0.270	0.02	14	0.14
			Antennal gland		0.110	0.12	21	0.58
Vertebrates								
Aquatic								
Tilapia	4	SW	Gill	37	0.495	6.79	22	30.14
	4	FW			0.356	4.71	19	24.49
	5		Gut		0.027	0.63	29	2.17
Trout	6	FW	Gill	37	0.160	1.86	nd	nd
Eel	7	FW	Gill	37	0.053	2.25	33	6.82
Terrestrial								
Rat	8	Terrestrial	Kidney	37	0.070	7.40	12	61.42
				25	0.130	4.30	12	35.69
	9		Duodenum	25	0.200	5.30	20	26.50
	10		Liver	37	0.014	30.00	15	200.00

nd, not determined; IOVs, inside out vesicles; SW, seawater; FW, fresh water
References: 1, Flik et al. (1994); 2, Zhuang & Ahearn (1998); 3, Wheatly et al. (1999); 4, Verbost et al. (1994); 5, Flik et al. (1990); 6, Perry & Flik (1988); Flik et al. (1985); 8, van Heeswijk et al. (1984); 9, Ghijsen et al. (1982); 10, Kraus-Friedmann et al. (1982).

the mitochondrial Ca^{2+}-ATPase inhibitors oligomycin and sodium azide, and thapsigargin, a SERCA inhibitor.

Addition of A23187 typically produces efflux of $^{45}Ca^{2+}$ preloaded into vesicles, proving that uptake occurs into the intravesicular space against a concentration gradient. In most vesicles ionophore-stimulated Ca^{2+} efflux is incomplete (70% over 10 min in crayfish, lobster hepatopancreas, rat kidney, eel gill, trout gill). Complete unloading (as in crab gill, rat liver and duodenum) implicates negligible Ca^{2+} binding in the interior of the vesicle. Typically, uptake is completely inhibited when A23187 is added from the start of the experiment (crayfish and trout vesicles); partial inhibition of uptake (rat enterocytes and kidney, eel gill) suggests that some Ca^{2+} entering via PMCA remains internally bound and thus does not equilibrate.

PMCA affinity, K_m, is typically between 0.01 and 0.5 μM, corresponding closely to the free $[Ca^{2+}]_i$ (100-200 nM). In crustaceans, K_m is similar irrespective of osmotic origin and does not appear to change upon external dilution (G Flik and C Haond, unpublished results). In fish, however, affinity tends to be higher in fresh water species. Similarly, a relatively low K_m is observed in terrestrial species commensurate with evolution on to land from fresh water origins.

The maximal flux rate, J_{max}, reflects pump density and transport capacity (calculated by correcting J_{max} to 100% IOVs) reflecting the role of each epithelium in calcium homeostasis. Within a tissue type, J_{max} tends to be up to 3-fold higher in species of seawater origin than in fresh water species (compare crustaceans and fish), commensurate with increased whole organism ion fluxes. Species that have evolved in fresh water typically have tight epithelia with minimal paracellular ion fluxes. In tilapia, pump density is comparable in fresh water and seawater, so calcium flux is governed by the permeability/transepithelial potential. Flik & Haond (unpublished results) found that PMCA transport capacity of lobster epipodites and branchiostegites increased in dilute seawater to compensate for limited ion regulatory capacity of the gills (where pump levels declined). Crustacean pump activity is elevated at times of mineralization (Wheatly 1999).

Similarly J_{max} can be compared among different epithelial tissues within an organism. For example, in both fresh water crayfish and fish, the gill transport capacity significantly exceeds that of the gut, suggesting that aquatic species rely more on external water for Ca^{2+} than on ingested materials. Gill Ca^{2+} uptake is more important in fresh water fish than in fresh water crayfish as fish mineralize continuously, whereas crustaceans mineralize their cuticle intermittently (primarily in postmolt). Evolution into terrestrial environments and loss of contact with water results in organisms becoming increasingly dependent upon gastric epithelia for Ca^{2+} influx from food, with the renal reabsorption of Ca^{2+} in urinary filtrate also contributing to Ca^{2+} homeostasis.

Temperature dependence of PMCA is of physiological significance to aquatic species that tend to be poikilothermic. Reported Q_{10} values for PMCA are comparable among all species (1.5-2.1; crayfish, rat kidney and pancreas) irrespective of environment.

The high affinity PMCA is typically, although not always, calmodulin dependent. EGTA treatment reduces ATP-dependent $^{45}Ca^{2+}$ uptake by chelating Ca^{2+} and thereby removing endogenous calmodulin (crayfish hepatopancreas and antennal gland, eel and trout gill). Any remaining uptake reflects basal calmodulin-independent PMCA activity which is significant in some vesicles (50% in crustacean hepatopancreas, fish gill, rat kidney) but negligible in others (10-20%, crayfish antennal gland, rat intestine). Calmodulin repletion restores ATP-dependent $^{45}Ca^{2+}$ uptake in EGTA-treated vesicles. Calmodulin exerts its effect via both J_{max} and K_m (rat duodenum and kidney). Calmodulin sensitivity could not be demonstrated in rat liver or in crayfish gill.

Introduction of extravesicular Na^+ (5mM) produces an ouabain-inhibitable increase in $^{45}Ca^{2+}$ uptake attributable to the Ca^{2+}/Na^+ exchanger. Under routine conditions ($[Ca^{2+}]_i$ less than 500 nM) the PMCA pump accounts for 80-90% of cellular Ca^{2+} efflux (intermolt crustacean hepatopancreas and antennal gland, mammalian intestine and kidney). At higher $[Ca]_i$, as may occur during rapid transepithelial calcium flux in certain molting phases, the exchanger may assume a more prominent role. In fish enterocytes, however, the Ca^{2+}/Na^+ exchanger had 6-fold less Ca^{2+} transport capacity than the pump.

When *in vitro* uptake capacity by an epithelium was compared with *in vivo* rates on a per animal basis, it became apparent that gills of fresh water species are engineered with an overcapacity to pump calcium (six times the capacity in crayfish, 200 times in tilapia) allowing for activity to be upregulated as needed (e.g. crustacean postmolt). The crayfish antennal gland, meanwhile, pumps at capacity during intermolt. In contrast, in the marine crab gill, calcium pumps are working at capacity in intermolt; postmolt upregulation is not necessary as Ca^{2+} diffuses into the postmolt crab passively.

SERCA has been characterized in purified sarcoplasmic reticulum. SERCA tends to have a lower affinity (2 µM) but higher capacity (50-400 nmol/min per mg protein) than PMCA, commensurate with its relative abundance (Simonides & van Hardeveld 1990, Wheatly 1999).

Molecular characterization of the calcium pump

SERCA and PMCA are both 'P'-type ATPases; cloning and sequencing of SERCA preceded that of PMCA because of its high abundance in tissues such as muscle. SERCA has approximately 1000 amino acid residues with three cytoplasmic domains joined to a set of 10 transmembrane helices by a narrow pentahelical stalk. ATP bound to one cytoplasmic domain would phosphorylate an aspartate in an adjoining domain inducing translocation of calcium from binding sites on the stalk. A summary of genes and splicing variants is given in Table 2.

In vertebrates, SERCA is encoded by three genes with five isoforms generated by differential processing of the transcripts that exhibit tissue-specific expression. SERCA1 and 2 are 84% homologous and there is 76% identity between amino acid sequences of SERCA1, 2 and 3. Vertebrate SERCA has a molecular mass of 110 kDa and a stoichiometry of 2 Ca^{2+} per ATP. SERCA1 is regulated by sarcolipin and

Calcium Metabolism: Comparative Endocrinology

Table 2 Calcium pump isoforms deduced from cDNA data

Isoform	Splicing Variants	Species (Reference)	Tissue Distribution (Relative expression)	Residues
SERCA				
Vertebrates				
1	a	Rabbit (1), rat (2), chicken (3), frog (4)	Adult fast muscle	997
	b	Rabbit, rat, chicken, frog	Neonatal fast muscle	1001
2	a	Rabbit, rat human, (5)	Slow and cardiac muscle	997
	b	Rabbit, rat, human	Ubiquitous	1043
3		Rat, human	Non-muscle	999
Invertebrates				
	4.5 kb	Brine shrimp(6),	Muscle (high), increases in development	1003
		crayfish (7)	Tail muscle (high) egg, gill, antennal gland, hepatopancreas (low)	1002
	5.2 kb	Brine shrimp	Ubiquitous, decreases in development	1027
	5.8 kb	Crayfish	Cardiac muscle (high), hepatopancreas, eggs (low)	1020
	7.6 kb	Crayfish	Egg (high), heart, gill, antennal gland, hepatopancreas (low)	
	8.8 kb	Crayfish	Heart	
	10.1 kb	Crayfish	Gill, antennal gland, hepatopancreas	
	3.3 kb	Fruit fly (8)	Nervous, muscle (high), all tissues (low)	1002
PMCA				
Vertebrates				
1	a-e	Rat, human (9, 10)	Ubiquitous, developmentally regulated	1171-1258
2	b, f	Rat, human	Brain, developmentally regulated	1198-1212
3	a	Rat	Brain, downregulated in adult	1159
4	a, b, g	Human	Ubiquitous, developmentally regulated	1169-1205
Invertebrates				
	5.4 kb	Crayfish (11)	Eggs only	N/A
	7.5 kb	Crayfish	Antennal gland (high), ubiquitous	

References: 1, Brandl et al. (1987); 2, Wu et al. (1995); 3, Campbell et al. (1991); 4, Vilsen & Anderson (1992); 5, Lytton & MacLennan (1988); 6, Palmero & Sastre (1989); 7, Z Zhang, D Chen & M Wheatly, unpublished; 8, Magyar et al. (1995); 9, Strehler (1991); 10, Carafoli (1992); 11, Z Zhang, F Castellano & M Wheatly, unpublished.
N/A, not available.

SERCA2 is regulated by phospholamban (increases J_{max}, decreases K_m) through calmodulin-dependent protein kinase and protein kinase A. SERCA2 is translated into two distinct proteins that differ in the C-terminus. One has a short tail of four amino acids; the longer version has 30-40 additional hydrophobic residues that have the potential of forming an extra transmembrane domain.

In invertebrates (specifically arthropods), SERCA is encoded by a single gene that most closely resembles vertebrate SERCA2 (70% amino acid sequence identity). Researchers have postulated the existence of a unique ancestral gene that gave rise to a single arthropod gene and three vertebrate genes. In brine shrimp and crayfish there are two prominent splicing variants (4.5, 5.8/5.2 kb) with different C-terminal extensions (either six to nine residues in the short tail, or 29-30 hydrophobic residues in the long tail) that share hydrophobic character but no significant homology with SERCA2 variants in vertebrates. In crayfish three additional gene products have been identified. The invertebrate gene exhibits high abundance in certain tissues consistent with a specific role in calcium homeostasis (muscle, contraction relaxation; nerve, signal transduction); low abundance in all other tissues implicates a housekeeping function. Prominence of this gene product in oocytes confirms its importance in embryogenesis; it appears to exhibit developmental regulation. In fruit fly a single transcript (3.3 kb) has been reported, suggesting that alternative splicing is not involved. The deduced amino acid sequences for SERCA in the three invertebrates studied share 70-83% identity.

In vertebrates, PMCA is encoded by four genes with 20 alternatively spliced gene products differing mainly in their affinity for calmodulin. Vertebrate PMCA has a molecular mass of 130-140 kDa, a stoichiometry of 1 Ca^{2+} per ATP, and can be regulated by calmodulin (increases J_{max}, decreases K_m) or acidic phospholipids and long-chain polyunsaturated fatty acids. Phosphorylation by protein kinase A increases J_{max} and decreases K_m, and phosphorylation by protein kinase C increases J_{max}. We have cloned a partial (1596 bp) cDNA fragment for crayfish antennal gland PMCA that corresponds to 45% of the coding region. The deduced amino acid sequence shares 60-70% homology with mammalian PMCA. Work to date suggests that there are two transcripts. One (5.4 kb) is found only in eggs and appears to be developmentally regulated; the other (7.5 kb) form is ubiquitous.

Vertebrate SERCA genes are differentially expressed during development and in response to treatments that alter calcium pumping activity such as chronic muscle stimulation. Our work has demonstrated that both SERCA and PMCA are differentially expressed during the molting cycle of crayfish, commensurate with changes in transepithelial calcium flux. SERCA expression is greatest in intermolt and decreases in postmolt, while PMCA expression (gills) is the opposite (least in intermolt, maximal in postmolt), suggesting that expression of SERCA and PMCA is inversely regulated in invertebrates as has been shown in mammals.

Perspectives

In the future we plan to study the regulation of calcium pump expression by the steroid ecdysone, a hormone that co-ordinates ecdysis. Ecdysone typically binds to an intranuclear DNA receptor. This complex then binds to an ecdysone response element in the gene promoter where it regulates transcription. We are also exploring the evolutionary relationships among PMCAs and SERCAs, and the physiological significance of isoform diversity, abundance and tissue distribution.

References

Brandl CJ, deLeon S, Martin DR & MacLennan DH (1987) Adult forms of the Ca^{2+}ATPase of sarcoplasmic reticulum. *Journal of Biological Chemistry* **262** 3768-3774.

Campbell A, Kessler P, Sagare Y, Inesi G & Fambrough D (1991) Nucleotide sequence of avian cardiac and brain SR/ER Ca^{2+} ATPase and functional comparisons with fast twitch Ca^{2+} ATPase: calcium affinities and inhibitor effects. *Journal of Biological Chemistry* **266** 16050-16055.

Carafoli E (1992) The Ca^{2+} pump of the plasma membrane. *Journal of Biological Chemistry* **267** 2115-2118.

Flik G, Wendelaar Bonga SE & Fenwick JC (1985) Active Ca^{2+} transport in plasma membranes of branchial epithelium of the North-American eel, *Anguilla rostrata* LeSueur. *Biology of the Cell* **55** 265-272.

Flik G, Schoenmakers TJM, Groot JA, Van Os CH & Wendelaar Bonga SE (1990) Calcium absorption by fish intestine: the involvement of ATP-and sodium-dependent calcium extrusion mechanisms. *Journal of Membrane Biology* **113** 13-22.

Flik G, Verbost PM, Atsma W & Lucu C (1994) Calcium transport in gill plasma membranes of the crab *Carcinus maenas*: evidence for carriers driven by ATP and a Na^+ gradient. *Journal of Experimental Biology* **195** 109-122.

Ghijsen WEJM, De Jong HR & van Os CH (1982) ATP-dependent calcium transport and its correlation with Ca^{2+}-ATPase activity in basolateral plasma membranes of rat duodenum. *Biochimica et Biophysica Acta* **689** 327-336.

Kraus-Friedmann N, Biber J, Murer H & Carafoli E (1982) Calcium uptake in isolated hepatic plasma-membrane vesicles. *European Journal of Biochemistry* **129** 7-12.

Lytton J & MacLennan DH (1988) Molecular cloning of cDNAs from human kidney coding for two alternatively spliced products of the cardiac Ca^{2+} ATPase gene. *Journal of Biological Chemistry* **263** 15024-15031.

Magyar A, Bakos E & Varadi A (1995) Structure and tissue-specific expression of the *Drosophila melanogaster* organellar-type Ca^{2+} ATPase gene. *Biochemical Journal* **310** 757-763.

Palmero I & Sastre L (1989) Complementary DNA cloning of a protein highly homologous to mammalian sarcoplasmic reticulum Ca-ATPase from the crustacean *Artemia*. *Journal of Molecular Biology* **210** 737-748.

Perry SF & Flik G (1988) Characterization of branchial transepithelial calcium fluxes in freshwater trout, *Salmo gairdneri*. *American Journal of Physiology* **254** R491-498.

Simonides W & van Hardeveld C (1990) An assay for sarcoplasmic reticulum Ca^{2+} ATPase activity in muscle homogenates. *Analytical Biochemistry* **191** 321-331.

Strehler EE (1991) Recent advances in the molecular characterization of the plasma membrane Ca^{2+} ATPase. *Journal of Membrane Biology* **20** 1-15.

Telford WG & Miller RA (1996) Detection of plasma membrane Ca^{2+}-ATPase activity in mouse T lymphocytes by flow cytometry using fluo-3-loaded vesicles. *Cytometry* **24** 243-250.

van Heeswijk MPE, Geertsen JAM & van Os CH (1984) Kinetic properties of the ATP-dependent Ca^{2+} pump and the Na^+/Ca^{2+} exchange system in basolateral membranes from rat kidney cortex. *Journal of Membrane Biology* **79** 19-31.

Verbost PM, Schoenmakers TJM, Flik G & Wendelaar Bonga SE (1994) Kinetics of ATP- and Na^+-gradient driven Ca^{2+} transport in basolateral membranes from gills of freshwater- and seawater-adapted tilapia. *Journal of Experimental Biology* **186** 95-108.

Vilsen B & Anderson J (1992) Deduced amino acid sequence and E1-E2 equilibrium of the sarcoplasmic reticulum Ca^{2+}-ATPase of frog skeletal muscle: comparison with the Ca^{2+} ATPase of rabbit fast twitch muscle. *FEBS Letters* **306** 213-218.

Wheatly MG (1999) Calcium homeostasis in crustacea: the evolving role of branchial, renal, digestive and hypodermal epithelia. *Journal of Experimental Zoology* **283** (In Press).

Wheatly MG, Pence RC & Weil JR (1999) ATP-dependent calcium uptake into basolateral vesicles from transporting epithelia of intermolt crayfish. *American Journal of Physiology* **276** R566-R574.

Wu K-D, Lee W-S, Wey J, Bungard D & Lytton J (1995) Localization and quantification of endoplasmic reticulum Ca^{2+}-ATPase isoform transcripts. *American Journal of Physiology* **269** C775-784.

Zhuang Z & Ahearn GA (1998) Energized Ca^{2+} transport by hepatopancreatic basolateral plasma membranes of *Homarus americanus*. *Journal of Experimental Biology* **201** 211-220.

Formation of the calcified exoskeleton in the prawn, *Penaeus japonicus*

P Persson[1,2], T Ikeya[1] and T Watanabe[1]

[1]Ocean Research Institute, University of Tokyo, 1-15-1 Minamidai, Nakano, Tokyo 164, Japan and [2]Fish Endocrinology Laboratory, Department of Zoology, Göteborg University, Box 463, SE-405 30 Göteborg, Sweden

Organization of the crustacean exoskeleton

The crustacean exoskeleton is shed during molting to permit growth, and a new exoskeleton is subsequently produced and calcified. Although the synthesis of the new exoskeleton is already initiated during the premolt stage, calcification of the newly synthesized exoskeleton takes place exclusively during the postmolt stage. During the intermolt period, the period between exoskeleton production and shedding, no growth, and thus no calcification of the exoskeleton occurs.

The crustacean exoskeleton is composed of four layers: the outermost epicuticle and, in order beneath, the exocuticle, the endocuticle and the membranous layer (Roer & Dillaman 1984, Simkiss & Wilbur 1989). The epicuticle matrix consists of lipoprotein, while the matrices of the other three layers are composed of chitin and protein. All layers, with the exception of the membranous layer, are calcified with $CaCO_3$. The exoskeleton is produced by epithelial cells in the hypodermis which is present underneath, and in contact with, the membranous layer (Simkiss & Wilbur 1989). These cells have long cytoplasmic extensions, which penetrate the four layers of the exoskeleton through pore canals, and terminate at or in the epicuticle (Roer & Dillaman 1984, Simkiss & Wilbur 1989). The pore canals are extremely numerous and all parts of the exoskeleton are thus in close contact with the hypodermis and should be considered living tissue (Roer & Dillaman 1984).

Formation of the crustacean exoskeleton

The organic matrix of the epicuticle and the exocuticle is laid down during the premolt stage, while the production of the two innermost layers as well as calcification of the exoskeleton occur after molting (Roer & Dillaman 1984, Simkiss & Wilbur 1989). Concomitant with the formation of the two outermost layers during the premolt stage is the partial resorption of both the mineral and the organic portions of the old exoskeleton (Roer & Dillaman 1984). After the old exoskeleton has been shed, calcification of the epicuticle and the exocuticle begins in the most external regions and proceeds proximally (Roer & Dillaman 1984) and mineral has been suggested to reach the outer portions of the exoskeleton by the pore canals. The formation of the endocuticle matrix is initiated during the postmolt stage, and calcification is concomitant with matrix formation (Roer & Dillaman 1984). The end of the postmolt stage is marked by the deposition of the membranous layer and the cessation of calcium deposition in the exoskeleton (Roer & Dillaman 1984).

Mechanisms of calcification
Calcification mechanisms are present in both vertebrates and invertebrates although certain differences exist due to the different chemical compositions of the hard tissues. However, in both vertebrate and invertebrate hard tissue, acidic, non-collagenous extracellular matrix proteins have been identified, and it is believed that these proteins may play an important role in calcification. In later vertebrates, several acidic non-collagenous proteins such as osteocalcin, osteonectin, osteopontin, bone sialoprotein and phosphophoryn have been isolated and characterized, and their ability to bind hydroxyapatite, to nucleate calcium phosphate and to regulate the subsequent crystal growth has received considerable attention (Oldberg et al. 1986, Romberg et al. 1986, Bolander et al. 1988, Linde et al. 1989, Fisher et al. 1990, Moore et al. 1991, Hunter & Goldberg 1993, Ritchie & Wang 1996, Fujisawa et al. 1996, Wada et al. 1996, MacDougall et al. 1997). Although their role in the calcification process in vivo remains to be fully elucidated, it is generally accepted that several of these proteins have stimulatory and/or inhibitory effects on nucleation of calcium phosphate crystals and the subsequent crystal growth. The currently accepted model of calcification in vertebrate hard tissue involves the promotion of calcium phosphate nucleation by acidic non-collagenous extracellular matrix proteins present in, and probably bound to, the collagenous extracellular framework. As calcification proceeds, the growth and structure of the hydroxyapatite crystals are regulated by the same type of proteins.

Less information is available on the acidic matrix proteins present in invertebrate hard tissue. A partial characterization of the soluble protein fraction of the exoskeleton from several mollusc species reveals that a significant portion is composed of a repeating sequence of aspartic acid separated by serine or glycine (Weiner & Hood 1975). The ability of some partially purified acidic matrix proteins to affect $CaCO_3$ nucleation and subsequent crystal growth has been investigated, and the results are consistent with that of the acidic non-collagenous matrix proteins in vertebrate hard tissue (Addadi & Weiner 1985, Belcher et al. 1996, Falini et al. 1996). However, in only a few cases have the amino acid sequence of invertebrate hard tissue matrix proteins been determined (Katoh-Fukui et al. 1991, Miyamoto et al. 1996, Sudo et al. 1997, Ishii et al. 1998). In the fresh water crayfish, *Procambarus clarkii*, an insoluble matrix protein present in the gastroliths has been purified and found to be rich in glutamic acid/glutamine (Ishii et al. 1998). Further, nacrein, a soluble matrix protein present in the nacreous layer of the pearl oyster, *Pinctada fucata*, has been purified and cloned, and the deduced amino acid sequence reveals a carbonic anhydrase domain and a putative calcium-binding domain rich in acidic amino acids (Miyamoto et al. 1996). As mollusc shells consist of two different layers, it has been suggested that the type of $CaCO_3$ crystals present, i.e. calcite in the nacreous mother-of-pearl layer and aragonite in the prismatic layer, is controlled by the matrix proteins present in each respective layer. In support of this is the finding of two insoluble matrix proteins, one

Calcium Metabolism: Comparative Endocrinology

in the nacreous layer and the other in the prismatic layer (Sudo *et al.* 1997). Further, acidic macromolecules extracted from the nacreous and the prismatic shell layers induce aragonite and calcite formation respectively (Belcher *et al.* 1996, Falini *et al.* 1996). Thus, despite the limited information about individual proteins in invertebrate hard tissue, current data strongly suggest that the nucleation and growth of $CaCO_3$ crystals are affected by proteins present in the matrix.

Specific mRNA expression during calcification of the prawn exoskeleton

It is apparent that little is known about the matrix proteins in invertebrate hard tissue, particularly those affecting the nucleation of $CaCO_3$ and the subsequent crystal growth. This is partly due to the difficulty of isolating and purifying matrix proteins, especially insoluble proteins and proteins present in low abundance. Therefore, an alternative approach has been used by the present authors in which the gene expression was investigated in the crustacean exoskeleton during the different stages of the molt cycle. In order to specifically identify proteins important to the formation of the calcified crustacean exoskeleton, the differential display technique (Liang & Pardee 1992) has been used to identify mRNA expressed specifically during the synthesis of the kuruma prawn, *Penaeus japonicus*, exoskeleton (P Persson, T Ikeya & T Watanabe, unpublished results). Several PCR products have been amplified to a detectable level in postmolt, but not in intermolt and premolt exoskeleton, suggesting that several genes are expressed exclusively during the calcification of the prawn exoskeleton.

In order to identify and characterize the specifically expressed proteins, a cDNA library has been constructed and screened, and the mRNA identified during the differential display amplified, isolated as cDNA and the nucleotide sequence determined. The nucleotide sequences of four cDNAs (DD1, DD4, DD5 and DD9) have been elucidated, and the amino acid sequence of the encoded proteins deduced. A homology search in the PIR and Swiss Prot databases using the FASTA program (Pearson & Lipman 1988) has been performed to reveal potential homology to previously reported proteins.

The DD1 cDNA contains an open reading frame of 213 amino acids, and a homology search indicated that this cDNA species may encode a protein not previously described. A putative signal sequence is present at the N-terminus of the protein (Nielsen *et al.* 1997), indicating that the protein is secreted from the cell.

The nucleotide sequence of the DD4 cDNA has been determined with the exception of its 5' end, and a partial open reading frame deduced. The encoded protein has some similarity to calphotin, a calcium-binding protein present in fruit fly photoreceptor cells (Balinger *et al.* 1993, Martin *et al.* 1993). The DD4 protein is relatively rich in proline and acidic amino acids, a common feature of proteins affecting the nucleation and growth of calcium carbonate/phosphate crystals. Preliminary calcium-binding studies using the radioactive isotope ^{45}Ca imply that the DD4 protein may have some calcium-binding ability. However, further experiments are needed to determine the ability of the protein to stimulate/inhibit the nucleation of $CaCO_3$ and to regulate the subsequent crystal growth.

Both the DD5 and DD9 protein share notable homology with small insect cuticular proteins (Andersen et al. 1995). A stretch of 28 hydrophilic amino acids, as well as a proline-rich region of eight amino acids, appears to be highly conserved within this group of proteins, indicating common ancestry (Andersen et al. 1995). Supporting this is the observation that several proteins extracted from the Bermuda land crab, *Gecarcinus lateralis*, cross-reacted with antibodies raised against insect cuticular proteins (Kumari & Skinner 1993). Further, the amino acid sequence of exoskeletal proteins purified from the Bermuda land crab has been partially determined and found to have motifs present also in insect cuticular proteins (Kumari et al. 1995).

Concluding remarks

Available data suggest that the nucleation of $CaCO_3$ and subsequent crystal growth in invertebrate hard tissue is governed by proteins present in the hard tissue matrix. As a first step to identify such proteins in the prawn exoskeleton, four proteins (DD1, DD4, DD5 and DD9) which are expressed almost exclusively during the postmolt stage, the period when the exoskeleton is calcified, have been identified. Thus, it is likely that these proteins play a central role in the formation of the calcified prawn exoskeleton, although their precise function remains to be elucidated.

References

Addadi L & Weiner S (1985) Interactions between acidic proteins and crystals: stereochemical requirements in biomineralization. *Proceedings of the National Academy of Sciences USA* **82** 4110-4114.

Andersen SO, Hojrup P & Roepstorff P (1995) Insect cuticular proteins. *Insect Biochemistry and Molecular Biology* **25** 153-176.

Ballinger DG, Xue N & Harsman KD (1993) A *Drosophila* photoreceptor cell-specific protein, calphotin, binds calcium and contains a leucine zipper. *Proceedings of the National Academy of Sciences USA* **90** 1536-1540.

Belcher AM, Wu XH, Christensen RJ, Hansma PK, Stucky GD & Morse DE (1996) Control of crystal phase switching and orientation by soluble mollusk-shell proteins. *Nature* **381** 56-58.

Bolander ME, Young MF, Fisher LW, Yamada Y & Termine JD (1988) Osteonectin cDNA sequence reveals potential binding regions for calcium and hydroxyapatite and shows homologies with both a basement membrane protein (SPARC) and a serine proteinase inhibitor (ovomucoid). *Proceedings of the National Academy of Sciences USA* **85** 2919-2923.

Falini G, Albeck S, Weiner S & Addadi L (1996) Control of aragonite or calcite polymorphism by mollusk shell macromolecules. *Science* **271** 67-69.

Fisher LW, McBride OW, Termine JD & Young MF (1990) Human bone sialoprotein. Deduced sequence and chromosomal localization. *Journal of Biological Chemistry* **265** 2347-2351.

Fujisawa R, Wada Y, Nodasaka Y & Kuboki Y (1996) Acidic amino acid-rich sequences as binding sites of osteonectin to hydroxyapatite crystals. *Biochimica et Biophysica Acta* **1292** 53-60.

Hunter GK & Goldberg HA (1993) Nucleation of hydroxyapatite by bone sialoprotein. *Proceedings of the National Academy of Sciences USA* **90** 8562-8565.

Ishii K, Tsutsui N, Watanabe T, Yanagisawa T & Nagasawa H (1998) Solubilization and chemical characterization of an insoluble matrix protein in the gastroliths of a crayfish, *Procambarus clarkii*. *Bioscience, Biotechnology and Biochemistry* **62** 291-296.

Katoh-Fukui Y, Noce T, Ueda T, Fujiwara Y, Hashimoto N, Higashinakagawa T, *et al.* (1991) The corrected structure of the SM50 spicule matrix protein of *Strongylocentrotus purpuratus*. *Developmental Biology* **145** 201-202.

Kurami SS & Skinner DM (1993) Proteins of crustacean exoskeleton II: immunological evidence for their relatedness to cuticular proteins of two insects. *Journal of Experimental Biology* **265** 195-210.

Kumari SS, Willis JH & Skinner DM (1995) Proteins of crustacean exoskeleton IV: partial amino acid sequences of exoskeletal proteins from the Bermuda land crab, *Gecarcinus lateralis*, and comparisons to certain insect proteins. *Journal of Experimental Biology* **273** 389-400.

Liang P & Pardee AB (1992) Differential display of eukaryotic messenger RNA by means of the polymerase chain reaction. *Science* **257** 967-971.

Linde A, Lussi A & Crenshaw MA (1989) Mineral induction by immobilized polyanionic proteins. *Calcified Tissue International* **44** 286-295.

MacDougall M, Simmons D, Luan X, Nydegger J, Feng J & Gu TT (1997) Dentin phosphoprotein and dentin sialoprotein are cleavage products expressed from a single transcript coded by a gene on human chromosome 4. *Journal of Biological Chemistry* **272** 835-842.

Martin JH, Benzer S, Rudnicka M & Miller CA (1993) Calphotin: a *Drosophila* photoreceptor cell calcium-binding protein. *Proceedings of the National Academy of Sciences USA* **90** 1531-1535.

Miyamoto H, Miyashita T, Okushima M, Nakano S, Morita T & Matsuhiro A (1996) A carbonic anhydrase from the nacreous layer in oyster pearls. *Proceedings of the National Academy of Sciences USA* **93** 9657-9660.

Moore MA, Gotoh Y, Rafidi K & Gerstenfeld LC (1991) Characterization of a cDNA for chicken osteopontin: expression during bone development, osteoblast differentiation, and tissue distribution. *Biochemistry* **30** 2501-2508.

Nielsen H, Engelbrecht J, Bunak S & von Heijne G (1997) Identification of prokaryotic and eukaryotic signal peptides and prediction of their cleavage sites. *Protein Engineering* **10** 1-6.

Oldberg Å, Franzen A & Heinegård D (1986) Cloning and sequence analysis of rat bone sialoprotein (osteopontin) cDNA reveals an Arg-Gly-Asp cell-binding sequence. *Proceedings of the National Academy of Sciences USA* **83** 8819-8823.

Pearson WR & Lipman DJ (1988) Improved tools for biological sequence comparison. *Proceedings of the National Academy of Sciences USA* **85** 2444-2448.

Ritchie HH & Wang LH (1996) Sequence determination of an extremely acidic rat dentin phosphoprotein. *Journal of Biological Chemistry* **271** 21695-21698.

Roer RD & Dillaman R (1984) The structure and calcification of the crustacean cuticle. *American Zoologist* **24** 893-909.

Romberg RW, Werness PG, Riggs BL & Mann KG (1986) Inhibition of hydroxyapatite crystal growth by bone-specific and other calcium-binding proteins. *Biochemistry* **25** 1176-1180.

Simkiss K & Wilbur KM (1989) *Biomineralization. Cell Biology and Mineral Deposition*. San Diego: Academic Press.

Sudo S, Fujikawa T, Nagakura T, Ohkubo T, Sakaguchi K, Tanaka M, *et al* . (1997) Structures of mollusk shell framework proteins. *Nature* **387** 563-564.

Wada Y, Fujisawa R, Nodasaka Y & Kuboki Y (1996) Electrophoretic gels of dentin matrix proteins as diffusion media for *in vitro* mineralization. *Journal of Dental Research* **75** 1381-1387.

Weiner S & Hood L (1975) Soluble protein of the organic matrix of mollusk shells: a potential template for shell formation. *Science* **190** 987-988.

Part Two

Fish

Calcium balance in teleost fish: transport and endocrine control mechanisms

B Th Björnsson, P Persson, D Larsson, S H Jóhannsson and K Sundell

Fish Endocrinology Laboratory, Department of Zoology, Göteborg University, Box 463, SE-405 30 Göteborg, Sweden

Introduction

Teleost species occupy both hypo- and hyper-calcemic environments and are thus normally subjected to a Ca gradient between the internal and external environment. In addition, migrating species may encounter rapid changes in salinity, temperature and acidity causing changes in drinking rate and food availability. These factors can rapidly affect the Ca-exchange rate between the fish and the environment, calling for equally rapid regulatory measures to maintain Ca homeostasis.

Research into teleost Ca regulation gained momentum with the pioneering work of Maurice Fontaine in the 1960s (Fontaine 1964) and became a major focus in comparative endocrinology during the 1970s and 1980s. Several comprehensive reviews have been compiled (Pang 1973, Dacke 1979, Clarke 1983, Hirano 1989, Wendelaar Bonga & Pang 1991, Flik & Verbost 1993, Flik *et al.* 1996), showing the endocrinology of the corpuscles of Stannius, together with the Ca-regulatory function of the gills, to have been a major research focus. Therefore, this review will be selective in highlighting aspects that have received less attention.

Ca transport and regulatory mechanisms
Gills
The chloride cells are believed to be the site of branchial Ca^{2+} uptake in marine as well as freshwater teleosts. Ca^{2+} entry across the brush-border membrane follows saturable kinetics and is believed to be mediated by Ca^{2+} channels and/or carriers (Flik & Verbost 1994). Inhibition of Ca^{2+} and Cd^{2+} entry into branchial Ca^{2+}-transporting cells by La^{3+} (Perry & Flik 1988, Wicklund Glynn *et al.* 1994, Comhaire *et al.* 1998) indicates that Ca^{2+} channels are present. The L-type Ca^{2+}-channel antagonists, verapamil and diltiazem, were found to be effective in decreasing branchial Cd^{2+} and Ca^{2+} uptake (Wicklund Glynn *et al.* 1994, Comhaire *et al.* 1998), but the L-type Ca^{2+}-channel antagonist, nifedipine, was without effect (Perry & Flik 1988). Thus, although somewhat contradictory, the results obtained show that L-type Ca^{2+} channels contribute to the apical Ca^{2+} entry into chloride cells, but other Ca^{2+}-transporting components are also indicated. Within the chloride cells, Ca^{2+} is either bound to Ca-binding proteins (CaBPs) or sequestered within intracellular organelles. A

branchial high-affinity CaBP was demonstrated in freshwater-adapted eel (*Anguilla rostrata*; Hearn et al. 1978). This 28 kDa protein has biochemical characteristics corresponding to those of calbindin, a protein found in Ca^{2+}-transporting cells of mammals and birds (Wasserman et al. 1992). The final step of Ca^{2+} uptake across chloride cells is an energy-dependent extrusion of Ca^{2+} across the basolateral membrane (BLM) mediated by Ca^{2+}-ATPases (Flik & Verbost 1993, 1994) and/or Na^+/Ca^{2+} exchangers (Flik & Verbost 1993, 1994, Verbost et al. 1994). There are some recent reviews covering this area (Flik et al. 1996, Marshall & Bryson 1998).

Branchial Ca^{2+} uptake is under endocrine control mainly by stanniocalcin (STC), which inhibits branchial Ca^{2+} uptake from the environment without affecting the efflux (Flik & Verbost 1993, Flik et al. 1996). The hormone inhibits Ca^{2+} entry into the chloride cells through Ca^{2+} channels or carriers via a cAMP-dependent pathway (Flik & Verbost 1994). A role for calcitonin (CT) in branchial Ca^{2+} regulation has been indicated, increasing efflux and decreasing influx in the eel (Milhaud et al. 1977, Milet et al. 1979). Branchial CT receptors have also been found in rainbow trout (Fouchereau-Peron et al. 1981, Arlot-Bonnemains et al. 1983), and CT gene expression has been found to take place in salmon gills (Martial et al. 1994). However, it appears that CT can have vasoconstrictory effects in isolated branchial preparations (Milhaud et al. 1977), which could account for some of the observed changes in branchial Ca^{2+} fluxes.

Intestine

In freshwater and marine fish, the intestine contributes at least 30% of the total body Ca^{2+} uptake (Sundell & Björnsson 1988, Flik et al. 1990). In the marine teleost, the Atlantic cod (*Gadus morhua*), intestinal Ca^{2+} absorption consists of a saturable (60%) and a non-saturable component (40%), reflecting that intestinal Ca^{2+} uptake takes place both through paracellular and transcellular pathways (Sundell & Björnsson 1988).

Intracellular Ca^{2+} concentrations are normally in the nanomolar range and the enterocyte interior is negatively charged compared with the intestinal lumen (Schoenmakers et al. 1992, Larsson et al. 1998). This allows Ca^{2+} to move across the brush-border membrane down an electrochemical gradient. In the freshwater tilapia (*Oreochromis mossambicus*), the Ca^{2+} uptake into isolated brush-border membrane vesicles is described as the sum of a saturable and a non-saturable component (Klaren et al. 1993). Intestinal Ca^{2+} uptake was further shown to be ATP dependent and suggested to be mediated by a P_2 purinoceptor-controlled Ca^{2+} channel or carrier in the enterocyte apical membrane (Klaren et al. 1997). In the marine Atlantic cod, Ca^{2+} entry was demonstrated to be through voltage-dependent L-type Ca^{2+} channels mainly localized in the brush-border region of the enterocytes (Larsson et al. 1998). After Ca^{2+} uptake into the enterocytes across the brush-border membrane, Ca^{2+} transport through the cytosol may be associated with proteins, as indicated by the presence of CaBPs in intestinal cells from the carp (Ooizumi et al. 1970, Chartier Baraduc 1973). There is also ample evidence for the presence of both Ca^{2+}-ATPases and/or Na^+/Ca^{2+} exchangers extruding Ca^{2+} from the cell in both freshwater and marine teleosts

(Flik et al. 1990, Schoenmakers & Flik 1992, Schoenmakers et al. 1993, D Larsson, H Lundqvist, BTh Björnsson, A Linde & K Sundell, unpublished results).

Endocrine regulation of intestinal Ca^{2+} transport involves STC as an inhibitor of Ca^{2+} uptake in freshwater-adapted rainbow trout and eel (Collie & Hirano 1984, Takagi et al. 1985) as well as in the marine cod (Sundell et al. 1992a). However, the main Ca^{2+}-regulatory hormones in the intestine are the vitamin D_3 metabolites. Both freshwater and marine fish possess cytosolic and/or nuclear intestinal receptors for 1,25-dihydroxyvitamin D_3 (1,25$(OH)_2D_3$) (Marcocci et al. 1982, Sundell et al. 1992b), and 1,25$(OH)_2D_3$ injections induce hypercalcemia after 24 h or longer (Sundell et al. 1993). Thus, a slow genome-mediated effect of 1,25$(OH)_2D_3$ seems to be present in both freshwater and marine fish (Sundell et al. 1996). A much more rapid non-genome-mediated effect of 1,25$(OH)_2D_3$ has also been found in freshwater eel, where it increased intestinal Ca^{2+} uptake within 10 min (Chartier et al. 1979), whereas such an effect was absent in the marine cod (Larsson et al. 1995). The presence of specific high-affinity binding of 1,25$(OH)_2D_3$ to BLMs of carp enterocytes, but not to the BLMs of cod enterocytes (D Larsson, I Nemere & K Sundell, unpublished results), further supports the hypothesis that there are fundamental differences in the regulatory role of the vitamin D_3 system between freshwater and marine teleost species.

Specific high-affinity membrane receptors for 24,25-dihydroxyvitamin D_3 (24,25$(OH)_2D_3$) have been found in BLMs from both carp and cod enterocytes (D Larsson, I Nemere & K Sundell unpublished results), and these have similar biochemical characteristics to those of the 24,25$(OH)_2D_3$ membrane receptors found in cockerel and rat (Nemere et al. 1994, 1998). In Atlantic cod enterocytes, 24,25$(OH)_2D_3$ inhibits Ca^{2+} influx dose-dependently by blocking L-type Ca^{2+} channels (Larsson et al. 1997, D Larsson & K Sundell, unpublished results). Thus, it appears that the long-term regulation of the enterocyte Ca^{2+} transporting machinery is governed by 1,25$(OH)_2D_3$ in both freshwater and marine fish, whereas the short-term minute-to-minute modulation of Ca^{2+}-uptake is mainly an increase by 1,25$(OH)_2D_3$ in freshwater fish and a decrease by 24,25$(OH)_2D_3$ in marine fish.

Kidney

In the Atlantic cod, Björnsson & Nilsson (1985) estimated the renal Ca^{2+} excretion to be about 70% of the total Ca^{2+} excretion, indicating renal Ca^{2+} secretion to be of major importance for marine teleosts, whereas freshwater species exhibit active renal Ca^{2+} reabsorption. In the euryhaline teleosts *Gillichthys mirabilis* (Doneen 1993) and *Oreochromis mossambicus* (Bijvelds et al. 1995), Ca^{2+} transport across the BLM was found to be energy dependent and mediated through high-affinity Ca^{2+}-ATPases, whereas there was no evidence for ouabain-sensitive or sodium-dependent Ca^{2+} transport (Bijvelds et al. 1995). In both *G. mirabilis* (Doneen 1993) and *O. mossambicus* (Bijvelds et al. 1995) there was a large decrease in the Ca^{2+}-ATPase activity when the fish were acclimated to seawater. This would result in increased Ca^{2+} excretion when the fish are in a hypercalcemic environment. The cytosolic Ca^{2+} transport in teleost kidney cells has only been studied in the freshwater-adapted eel (*A. rostrata*), where a 28 kDa high-affinity CaBP was found (Hearn et al. 1978), and the mechanisms of Ca^{2+} transport across the apical membrane of the ion-transporting kidney cells are not known.

Studies on the endocrine control of renal Ca^{2+} excretion and reabsorption have so far been restricted to the effects of removal of the corpuscles of Stannius in the eel (*A. rostrata*). Fenwick (1974) demonstrated an increased Ca^{2+} clearance after stannioectomy, and Butler & Alia Cadinouche (1995) found evidence to suggest that the increased renal Ca^{2+} excretion was due to increased total plasma Ca levels and not to changes in Ca^{2+} reabsorption.

Calcified tissues

Teleost fish is the earliest extant vertebrate group that possess true bone tissue. In addition to the calcified endoskeleton (bone), they also possess a calcified exoskeleton (scale). Teleost bone can both be cellular and acellular, the latter defined as lacking osteocytes (Moss 1961). Teleost scales are generally considered to be acellular, although in some species, cells have been observed within the fibrillary plate suggesting a certain cellularization (Meunier 1987).

Osteoblasts and osteoclasts, responsible for calcified tissue formation and resorption respectively, are found in scales as well as both cellular and acellular bone (Weiss & Watabe 1979, Sire *et al.* 1990, Takagi & Yamada 1992, Bereither-Hahn & Zylberberg 1993).

While normal somatic/skeletal growth will result in long-term net accretion of Ca into calcified tissues, gonadal growth during female sexual maturation may result in substantial transient Ca resorption from these tissues. Estradiol-17β (E_2)induces the liver to produce vitellogenin, which binds Ca^{2+} and is then sequestered by the oocytes, causing a substantially increased Ca demand by the female (Persson *et al.* 1998). This must be met by increased Ca^{2+} uptake from the environment and/or mobilization from internal Ca stores. Indeed, scale resorption increases in both sexes of several salmonid species during sexual maturation and spawning migration (Crichton 1935, Ouchi *et al.* 1972, Persson *et al.* 1998), and in female Atlantic salmon, *Salmo salar*, a concurrent increase in scale osteoclast activity is found during vitellogenesis (Persson *et al.* 1998). It thus appears that at least a part of the Ca^{2+} accumulating in the oocytes is being resorbed from scales (Persson *et al.* 1998). This is supported by studies showing that E_2 treatment induces Ca^{2+} mobilization from scales in rainbow trout (Carragher & Sumpter 1991, Persson *et al.* 1994), and that at least a part of this Ca^{2+} mobilization is due to increased scale osteoclast activity (Persson *et al.* 1995, 1997). This further suggests that E_2 has a role in the regulation of scale osteoclast activity in female salmonids. In contrast, E_2 treatment did not affect Ca^{2+} mobilization from, and decreased resorption of, rainbow trout bone (Carragher & Sumpter 1991, Persson *et al.* 1994, 1997). Thus, it may be hypothesized that the scales are resorbed, but the skeleton is protected by E_2 during sexual maturation, similar to its effects on the skeleton in later vertebrates (Gay *et al.* 1993, Brubaker & Gay 1994). There is also evidence that vitamin D_3 metabolites may protect the skeleton during sexual maturation. Although $1,25(OH)_2D_3$ increases bone demineralization in immature fish, an effect that can be counteracted by $24,25(OH)_2D_3$ which increases osteoblast activity (Lopez *et al.* 1980,

Wendelaar Bonga et al. 1983), it has been found to increase bone formation in sexually mature eel (Lopez et al. 1980). E_2 treatment also increases Ca^{2+} uptake from the surrounding water (Persson et al. 1994), and it is proposed that mobilizing Ca^{2+} from the environment is favored when sufficient Ca is available in the water and/or food. When the Ca demand exceeds the capacity of the Ca^{2+}-uptake mechanisms, or if it is not energetically favorable to extract Ca^{2+} from the environment, as appears to be the case during salmonid reproduction, the fish will mobilize Ca^{2+} from internal stores.

A high-affinity low-capacity E_2-binding component is present in rainbow trout scale, and E_2 receptor mRNA has been detected in both rainbow trout scale and bone (Armour et al. 1997, P Persson, JM Shrimpton, BTh Björnsson & SD McCormick, unpublished results), implying that E_2 acts directly on the scale cells. This is further confirmed by the fact that E_2 stimulates the scale osteoclast activity *in vitro* (Suzuki et al. 1998).

Concluding remarks

Although Ca regulation in teleosts has been extensively studied, we are still far from understanding this complex control system fully. Although the hormones discussed in this review as well as STC (Hirano 1989) are clearly involved in Ca regulation, their roles have still not been fully defined as the endocrine control of renal Ca handling, for example, remains to be elucidated. Other hormones such as parathyroid hormone-related protein (Trivett *et al* 1999, this volume) and somatolactin (Kaneko & Hirano 1993) may be involved in Ca regulation, and the search for parathyroid hormone in fish continues (Rubin and Jüppner 1999, this volume).

The direct and indirect effects on Ca balance of numerous osmoregulatory and/or growth-promoting hormones such as prolactin, cortisol, growth hormone, insulin-like growth factor I and thyroid hormones need also to be considered for a comprehensive understanding of the endocrine control.

It is of importance that studies on cellular Ca-transport mechanisms continue, as well as those on the recently identified Ca-sensing receptor (Ingleton *et al* 1999, this volume).

Much new knowledge has, however, been gained during recent years. Although it is difficult to generalize when only a few of the widely diversified teleost species have been studied, it is now recognized that Ca fluxes can be increased and decreased by antagonizing endocrine systems, and that teleost fish exhibit many of the same cellular transport mechanisms as later vertebrates. The roles of the intestine and the scales have been further clarified, and noteworthy evolutionary differences between freshwater and marine fish have been identified in terms of the regulatory mechanisms of the vitamin D_3 system. Such differences may also be revealed for other hormones when their cellular transport mechanisms are further studied.

It is clear that the international research effort to elucidate Ca regulation in teleosts has declined significantly in the 1990s. This is not due to the lack of challenging topics or able scientists, but primarily to an international shift in research funding policies, from basic to applied science. Thus, for research on this complex control system to regain momentum in the new millennium, funding agencies must refocus on the importance of basic science.

Acknowledgements
We thank our colleagues, Drs T Lundgren, H Lundqvist, I Nemere, A W Norman and Y Takagi for inspiring and fruitful collaborations, and the Swedish Natural Science Research Council for financial support.

References
Arlot-Bonnemains Y, Fouchereau-Peron M, Moukhtar MS & Milhaud G (1983) Characterization of target organs for calcitonin in lower and higher vertebrates. *Comparative Biochemistry and Physiology* **76A** 377-380.

Armour KJ, Lehane DB, Pakdel F, Valotaire Y, Russel R, Graham G & Henderson IW (1997) Estrogen receptor mRNA in mineralized tissues of rainbow trout: calcium mobilization by estrogen. *FEBS Letters* **411** 145-148.

Bereither-Hahn J & Zylberberg L (1993) Regeneration of teleost fish scale. *Comparative Biochemistry and Physiology* **105A** 625-641.

Bijvelds MJC, van der Heijden AJH, Flik G, Verbost PM, Kolar ZI & Wendelaar Bonga SE (1995) Calcium pump activities in the kidneys of *Oreochromis mossambicus*. *Journal of Experimental Biology* **198** 1351-1357.

Björnsson BTh & Nilsson S (1985) Renal and extra-renal excretion of calcium in the marine teleost, *Gadus morhua*. *American Journal of Physiology* **248** R18-R22.

Brubaker KD & Gay CV (1994) Specific binding of estrogen to osteoclast surfaces. *Biochemistry and Biophysics Research Communications* **200** 899-907.

Butler DG & Alia Cadinouche MZ (1995) Fractional reabsorption of calcium, magnesium and phosphate in the kidneys of freshwater North American eels (*Anguilla rostrata* LeSueur) following removal of the corpuscles of Stannius. *Journal of Comparative Physiology* **165B** 348-358.

Carragher JF & Sumpter JP (1991) The mobilization of calcium from calcified tissues of rainbow trout (*Oncorhynchus mykiss*) induced to synthesize vitellogenin. *Comparative Biochemistry and Physiology* **99A** 169-172.

Chartier Baraduc M-M (1973) Présence et thermostabilité de protéines liant le calcium dans les muqueuses intesinales et branchiales de divers Téléostéens. *Comptes Rendus de l'Academie des Sciences Paris* **276D** 785-788.

Chartier M-M, Millet C, Martelly E, Lopez E & Warrot S (1979) Stimulation par la vitamine D_3 et le 1,25-dihydroxyvitamine D_3 de l'absorption intestinale du calcium chez l'anguille (*Anguilla anguilla* L.). *Journal of Physiology Paris* **75** 275-282.

Collie NL & Hirano T (1984) Effects of Stanniectomy on intestinal calcium transport in the Japanese eel. *Zoological Science. Proceedings. 55th Annual Meeting of the Zoological Society of Japan* **1** 966.

Comhaire S, Blust R, VanGinneken L, Verbost P & Vanderborght O (1998) Branchial cobalt uptake in the carp, *Cyprinus carpio*: effect of calcium channel blockers and calcium injection. *Fish Physiology and Biochemistry* **18** 1-13.

Clarke NB (1983) Evolution of calcium regulation in lower vertebrates. *American Zoologist* **23** 719-727.

Crichton I (1935) Scale-absorption in salmon and sea trout. *Fisheries Board of Scotland, Salmon Fisheries* **4** 1-12.

Dacke CG (1979) *Calcium Regulation in Sub-Mammalian Vertebrates.* London: Academic Press.

Doneen BA (1993) High-affinity Ca^{2+}-Mg^{2+}-ATPase in kidney of euryhaline *Gillichthys mirabilis*: Kinetics, subcellular distribution and effects of salinity. *Comparative Biochemistry and Physiology* **106B** 719-728.

Fenwick JC (1974) The corpuscles of Stannius and calcium regulation in the North American eel (*Anguilla rostrata* LeSueur). *General and Comparative Endocrinology* **23** 127-135.

Flik G & Verbost PM (1993) Calcium transport in fish gills and intestine. *Journal of Experimental Biology* **184** 17-29.

Flik G & Verbost PM (1994) Ca^{2+} transport across plasma membranes. *Biochemistry and Molecular Biology of Fishes,* pp 625-637. Eds PW Hochachaka & TW Momsen. Amsterdam: Elsevier Science BV.

Flik G, Schoenmakers TJM, Groot JA, van Os CH & Wendelaar Bonga SE (1990) Calcium absorption by fish intestine: the involvement of ATP- and sodium-dependent calcium extrusion mechanisms. *Journal of Membrane Biology* **113** 13-22.

Flik G, Klaren PHM, Schoenmakers TJM, Bijvelds MJC, Verbost PM & Wendelaar Bonga SE (1996) Cellular calcium transport in fish: unique and universal mechanisms. *Physiological Zoology* **69** 403-417.

Fontaine M (1964) Corpuscules de Stannius et régulation ionique (Ca, K, Na) du milieu intérieur de l'anguille (*Anguilla anguilla* L.). *Comptes Rendus de l'Academie des Sciences Paris* **259** 875-878.

Fouchereau-Peron M, Mouktar MS, Le Gal Y & Milhaud G (1981) Demonstration of specific receptors for calcitonin in isolated trout gill cells. *Comparative Biochemistry and Physiology* **68A** 417-421.

Gay CV, Kief NL & Bekker PJ (1993) Effect of estrogen on acidification in osteoclasts. *Biochemistry and Biophysics Research Communications* **192** 1251-1259.

Hearn PR, Tomlinson S, Mellersh H, Preston CJ, Kenyon CJ & Russell RGG (1978) Low molecular weight calcium-binding protein from the kidney and the gill of the freshwater eel (*Anguilla anguilla*). *Journal of Endocrinology* **79** 36P-37P.

Hirano T (1989) The corpuscles of Stannius. *Vertebrate Endocrinology: Fundamentals and Biomedical Implications, Regulation of Calcium and Phosphate,* pp 139-169. Eds PKT Pang & MP Schreibman. San Diego: Academic Press Inc.

Ingleton PM *et al* (1999)The calcium-sensing receptor in fishes. *Calcium Metabolism: Comparative Endicrinology,* pp 45-48. Eds J Danks, C Dacke, G Flik & C Gay. Bristol: Bio Scientifica.

Kaneko T & Hirano T (1993) Role of prolactin and somatolactin in calcium regulation in fish. *Journal of Experimental Biology* **184** 31-45.

Klaren PHM, Flik G, Lock RAC & Wendelaar Bonga SE (1993) Ca^{2+} transport across intestinal brush border membranes of the cichlid teleost *Oreochromis mossambicus*. *Journal of Membrane Biology* **132** 157-166.

Klaren PH, Wendelaar Bonga SE & Flik G (1997) Evidence for P_2-purinoceptor-mediated uptake of Ca^{2+} across a fish (*Oreochromis mossambicus*) intestinal brush border membrane. *Biochemical Journal* **322** 29-34.

Larsson D, Björnsson BTh & Sundell K (1995) Physiological concentrations of 24,25-dihydroxyvitamin D_3 rapidly decrease the *in vitro* intestinal calcium uptake in the Atlantic cod, *Gadus morhua*. *General and Comparative Endocrinology* **100** 211-217.

Larsson D, Lundgren T & Sundell K (1997) Non-genomic actions of 25(OH)D$_3$, 24,25(OH)$_2$D$_3$ and 1,25(OH)$_2$D$_3$ on [Ca^{2+}]$_i$ in enterocytes from the Atlantic cod *(Gadus morhua)*. *Proceedings of the 10th Workshop on Vitamin D, Strasbourg, France*, pp 389-390. Eds AW Norman, R Bouillon & M Thomasset. Berlin: Walter de Gruyter Inc.

Larsson D, Lundgren T & Sundell K (1998) Ca^{2+} uptake through voltage-gated L-type Ca^{2+} channels by polarized enterocytes from Atlantic cod *Gadus morhua*. *Journal of Membrane Biology* **164** 229-237.

Lopez E, MacIntyre I, Martelly E, Lallier F & Vidal B (1980) Paradoxical effect of 1,25-dihydroxycholecalciferol on osteoblastic activity in the skeleton of the eel *Anguilla anguilla*. *Calcified Tissue Research* **22** 19-23.

Marcocci C, Freake HC, Iwasaki J, Lopez E & MacIntyre I (1982) Demonstration and organ distribution of the 1,25-dihydroxyvitamin D$_3$-binding protein in fish (*A. anguilla*). *Endocrinology* **110** 1347-1354.

Marshall WS & Bryson SE (1998) Transport mechanisms of seawater teleost chloride cells: an inclusive model of a multifunctional cell. *Comparative Biochemistry and Physiology* **119A** 97-106.

Martial K, Maubras L, Taboulet J, Julienne A, Berry M, Milhaud G, Bensson AA & Mouhktar MS (1994) The calcitonin gene is expressed in salmon gills. *Proceedings of the National Academy of Sciences USA* **91** 4912-4914.

Meunier FJ (1987) Os cellulaire, os acellulaire et tissues derives chez les osteichthyens: les phenomenes de l'acellularisation et de la perte de mineralisation. *Annals de Biologie Clinique (Paris)* **XXVI** 201-233.

Milet C & Peignoux-Deville JME (1979) Gill calcium fluxes in the eel *Anguilla anguilla* (L). Effects of Stannius corpuscles and ultimobranchial body. *Comparative Biochemistry and Physiology* **63A** 63-70.

Milhaud G, Rankin JC, Bolis L & Benson AA (1977) Calcitonin: its hormonal action on the gill. *Proceedings of the National Academy of Sciences of the USA* **74** 4693-4696.

Moss ML (1961) Osteogenesis of acellular teleost fish bone. *American Journal of Anatomy* **108** 99-110.

Nemere I, Dormanen MC, Hammond MW, Okamura WH & Norman AW (1994) Identification of a specific binding protein for 1α,25-dihydroxyvitamin D$_3$ in basal-lateral membranes of chick intestinal epithelium and relationship to transcaltachia. *Journal of Biological Chemistry.* **269** 23750-23756.

Nemere I, Schwartz Z, Pedrozo H, Sylvia VL, Dean DD & Boyan BD (1998) Identification of a membrane receptor for 1,25-dihydroxyvitamin D-3 which mediates rapid activation of protein kinase C. *Journal of Bone and Mineral Research* **13** 1353-1359.

Ooizumi K, Moriuchi S & Hosoya N (1970) Comparative study of vitamin-D$_3$-induced calcium binding protein. *Vitamins* **42** 171-175.

Ouchi K, Yamada J & Kosaka S (1972) On the resorption of scales and associated cells in the precocious male parr of the *masu* salmon (*Oncorhynchus masou*). *Bulletin of the Japanese Society of Scientific Fisheries* **38** 423-430.

Pang PKT (1973) Endocrine control of calcium metabolism in teleosts. *American Zoologist* **13** 775-792.

Perry SF & Flik G (1988) Characterization of branchial transepithelial calcium fluxes in freshwater trout, *Salmo gairdneri*. *American Journal of Physiology* **254** 491-498.

Persson P, Sundell K & Björnsson BTh (1994) Estradiol-17β-induced calcium uptake and resorption in juvenile rainbow trout, *Oncorhynchus mykiss*. *Fish Physiology and Biochemistry* **13** 379-86.

Persson P, Takagi Y & Björnsson BTh (1995) Tartrate resistant acid phosphatase as a marker for scale resorption in rainbow trout, *Oncorhynchus mykiss*: effects of estradiol-17β treatment and refeeding. *Fish Physiology and Biochemistry* **14** 329-339.

Persson P, Johannsson SH & Takagi Y & Björnsson BTh (1997) Estradiol-17β and nutritional status affect calcium balance, scale and bone resorption, and bone formation in rainbow trout, *Oncorhynchus mykiss*. *Journal of Comparative Physiology B* **167** 468-473.

Persson P, Sundell K, Björnsson BTh & Lundqvist H (1998) Calcium metabolism and osmoregulation during sexual maturation of river running Atlantic salmon (*Salmo salar* L.). *Journal of Fish Biology* **52** 334-349.

Rubin DA, & Jüppner H (1999) Molecular cloning of cDNAs encoding three distinct receptors for parathyroid hormone (PTH)/PTH-related peptide in the zebrafish. *Calcium Metabolism: Comparative Endicrinology*, pp 59-64. Eds J Danks, C Dacke, G Flik & C Gay. Bristol: Bio Scientifica.

Schoenmakers TJM & Flik G (1992) Sodium-extruding and calcium-extruding sodium/calcium exchangers display similar calcium affinities. *Journal of Experimental Biology* **168** 151-159.

Schoenmakers TJM, Klaren PHM, Flik G, Lock RAC, Pang PKT & Wendelaar Bonga SE (1992) Actions of cadmium on basolateral plasma membrane proteins involved in calcium uptake by fish intestine. *Journal of Membrane Biology* **127** 161-172.

Schoenmakers TJM, Verbost P, Flik G & Wendelaar Bonga SE (1993) Transcellular intestinal calcium transport in freshwater and seawater fish and its dependence on sodium/calcium exchange. *Journal of Experimental Biology* **176** 195-206.

Sire J, Huysseune A & Meunier FJ (1990) Osteoclasts in teleost fish: light- and electron-microscopical observations. *Cell and Tissue Research* 85-94.

Sundell K (1992) Intestinal calcium regulation in the marine teleost, *Gadus morhua*. Importance of stanniocalcin and the vitamin D_3 system. *PhD Thesis*, University of Göteborg.

Sundell K & Björnsson BTh (1988) Kinetics of calcium fluxes across the intestinal mucosa of the marine teleost, *Gadus morhua*, measured using an *in vitro* perfusion method. *Journal of Experimental Biology* **140** 171-86.

Sundell K, Björnsson BTh, Itoh H & Kawauchi H (1992*a*) Chum salmon (*Oncorhynchus keta*) Stanniocalcin inhibits *in vitro* intestinal calcium uptake in Atlantic cod (*Gadus morhua*). *Journal of Comparative Physiology* **162** 489-495.

Sundell K, Bishop JE, Björnsson BTh & Norman AW (1992*b*) 1,25-Dihydroxyvitamin D_3 in the Atlantic cod: plasma levels, a plasma binding component, and organ distribution of a high affinity receptor. *Endocrinology* **131** 2279-2286.

Sundell K, Norman AW & Björnsson BTh (1993) 1,25(OH)$_2$ vitamin D_3 increases ionized plasma calcium concentrations in the immature Atlantic cod *Gadus morhua*. *General and Comparative Endocrinology* **91** 344-351.

Sundell K, Larsson, D & Björnsson BTh (1996) The calcium regulatory role of the vitamin D_3-system in teleosts. *The Comparative Endocrinology of Calcium Regulation*, pp 75-84. Eds C Dacke, J Danks, I Caple & G Flik. Bristol: Journal of Endocrinology Ltd.

Suzuki N, Suzuki T & Kurokawa T (1998) Physiological role of calcitonin in fish: calcitonin suppresses osteoclastic activity in the scales of the goldfish (freshwater teleost) and the nibbler (seawater teleost). *Proceedings of 23rd Comparative Endocrinology Annual Conference, Kamaishi, Japan.*

Takagi Y & Yamada J (1992) Effects of calcium deprivation on the metabolism of acellular bone in tilapia, *Oreochromis niloticus*. *Comparative. Biochemistry and Physiology* **102A** 481-485.

Takagi Y, Nakamura Y & Yamada J (1985) Effects of the removal of corpuscles of Stannius on the transport of calcium across the intestine of rainbow trout. *Zoological Science* **2** 523-530.

Trivett MK *et al* (1999) Parathyroid hormone-related protein: localisation in cartilaginous fish tissues. *Calcium Metabolism: Comparative Endicrinology*, pp 49-58. Eds J Danks, C Dacke, G Flik & C Gay. Bristol: Bio Scientifica.

Verbost PM, Schoenmakers JM, Flik G & Wendelaar Bonga SE (1994) Kinetics of ATP- and Na^+-gradient driven Ca^{2+} transport in basolateral membranes from gills of freshwater- and seawater-adapted tilapia. *Journal of Experimental Biology* **186** 95-108.

Wasserman RH, Chandler JS & Meyer SA (1992) Intestinal calcium transport and calcium extrusion processes at the basolateral membrane. *Journal of Nutrition* **122** 662-671.

Weiss RE & Watabe N (1979) Studies on the biology of fish bone. III. Ultrastructure of osteogenesis and resorption in osteocytic (cellular) and anosteocytic (acellular) bones. *Calcified Tissue International* **28** 43-56.

Wendelaar Bonga SE & Pang PKT (1991) Control of calcium regulating hormones in the vertebrates: parathyroid hormone, calcitonin, prolactin and Stanniocalcin. *International Review of Cytology*, pp 139-212. Eds KW Jeon & M Friedland. San Diego: Academic Press Inc.

Wendelaar Bonga SE, Lammers PI & van der Meij JCA (1983) Effects of 1,25- and 24,25-dihydroxyvitamin D_3 on bone formation in the cichlid teleost *Sarotherodon mossambicus*. *Cell and Tissue Research* **228** 117-126.

Wicklund Glynn A, Norrgren L & Müssener Å (1994) Differences in uptake of inorganic mercury and cadmium in the gills of the zebrafish, *Brachydanio rerio*. *Aquatic Toxicology* **30** 13-29.

Osteocalcin response to environmental stressors

P E Patterson-Buckendahl[1], Z M G S Jahangir[1], M Rusnak[2] and R Kvetnansky[2]

[1]Division of Natural Sciences and Mathematics, Richard Stockton College, Pomona, New Jersey 08240, USA and [2]Institute of Experimental Endocrinology, Slovak Academy of Sciences, Bratislava, Slovakia

Introduction

Osteocalcin (OC), an abundant extracellular calcium-binding protein of bone origin, binds with high specificity to bone mineral and is remarkably stable in this form. A small but consistent portion is released into the circulation (pOC). Its protein sequence is at least 30% conserved across nearly 20 species representing five vertebrate classes from fish to mammals, especially in the mid region which contains an absolutely conserved sequence of three residues of 4-carboxyglutamic acid and a disulfide bridge (Hauschka et al. 1989). It has also been isolated from both bone and scales of *Lepomus macrochirus* (Nishimoto et al. 1992). The precise physiological functions of OC in bone or the circulation remain to be defined. In bone, it has been associated with both osteoclastic resorption (Glowacki & Lian 1987) and control of osteoblast function during formation (Ducy et al. 1996). pOC has been well correlated with histological indices of osteoblastic activity in endocrine diseases of bone (Delmas et al. 1985).

We previously showed that a variety of environmental stressors decreased rat pOC within 1.5 to 24 h regardless of age. These included cold exposure, noise, change of cage and other influences likely to produce mental anxiety. Administration of physiological levels of glucocorticoids replicated the pOC decline (Patterson-Buckendahl et al. 1988). A similar effect was reported in humans hospitalized for non-traumatic reasons (Napal et al. 1993). We subsequently investigated the effects of 2 h of foot restraint immobilization (IMMO), which models the fight-or-flight response necessary for survival in wild animals (Kvetnansky & Mikulaj 1970). In addition to the expected increases in epinephrine, norepinephrine and corticosterone, this more acute and severe stressor induced a rapid and dramatic increase in pOC within 5 min (Patterson-Buckendahl et al. 1995). Ablation of epinephrine by adrenalmedullectomy, norepinephrine by sympathetic neural blockade, and corticosterone by adrenalectomy showed that both glucocorticoid and sympathetic neural hormonal regulation were needed to restore pOC to baseline levels. Repeated imposition of IMMO daily for 42 times blunted but did not abolish the response and also resulted in decreased bone growth and a specific decrease in bone OC concentration (Patterson-Buckendahl et al. 1996). We do not know what the initial stimulus is for the rise in pOC, nor do we know whether its source is bone surface, increased secretion, or increased synthesis. We do know that it is accompanied by hypocalcemia and an increase in parathyroid hormone and that renal clearance is normal under IMMO (unpublished observations). These are consistent with the response of humans to citrate infusion (Gundberg *et al.* 1991) and that of monkeys anesthetized with isoflurane (Hotchkiss *et al.* 1998).

Because of the consistency of the response to acute stressors, we believe that pOC is an important part of the stress response, and may even be serving a hormonal role (Patterson-Buckendahl *et al.* 1996). If so, it should be present in all vertebrate species. We have therefore begun experiments to establish the presence of OC in species phylogenetically distant from mammals with the eventual goal of investigating the stress response in the most distant species. OC has been isolated and sequenced from three fish species: swordfish (Price *et al.* 1976), bluegill (Nishimoto *et al.* 1992) and sea bream (Cancela *et al.* 1995). A related protein, termed matrix Gla protein or MGP, was isolated from calcified shark cartilage; however, no detectable OC was present in that tissue (Rice *et al.* 1994). This raises a number of questions. (1) If pOC is important to normal function of vertebrates including sharks, and if they have no reservoir of the protein in bone, then do they exist without it, or is it synthesized in some other tissue such as the dermal denticles? (2) Some teleost fish mobilize mineral from scales under stressful conditions, and scales contain OC (Nishimoto *et al.* 1992). Does OC participate in that mobilization? (3) Elasmobranch dermal denticles are analogous to teeth, which in mammals contain OC in the dentine. Do dermal denticles contain OC and might it be a ready source of OC as in scales of teleost fish? We have approached these questions with both protein and genetic analysis.

Methods

Thresher shark (*Alopias vulpinus*) skin was obtained from a local supermarket. Skate (*Raja erinacea*) skin was from a fish caught in a census trawl off shore near Atlantic City and was already dead when obtained. Skin was cleaned of muscle and as much fascia as possible, cut into small strips and lyophilized. Dried strips were processed in a blender to break them up further, then extracted with 0.5 M EDTA containing protease inhibitors (Hauschka *et al.* 1989). This produced a very gelatinous extract which was dialyzed in a Spectrapor 3000 molecular mass cut-off membrane against ice-cold distilled water. After dialysis, the mixture was centrifuged and the supernatant decanted, diluted with water and passed over a bulk C_{18} reversed-phase column to desalt and remove much of the extraneous protein. The column was washed with water containing 0.1% trifluoroacetic acid (TFA) followed by 30% methanol in TFA, and eluted with 80% methanol in TFA. The eluate was dried, resuspended in 50 mM Tris buffer, pH 8.0 (Tris), and passed over a pre-equilibrated Sephadex G75 column. Fractions were collected, pooled, and loaded on to a Bio-Rad EconoQ column. Protein was eluted with a gradient of 0 to 0.6 M NaCl in Tris. Fractions were qualitatively assayed for carboxyglutamic acid content by reaction with diazobenzenesulfonic acid (Nishimoto 1990). Protein concentration was independently determined with BCA reagents (Pierce, Rockford, IL, USA).

Mouse OC cDNA probe was a gift from Dr Gerard Karsenty, Baylor University Medical Center, Houston, TX, USA. A sample of 1 µg of the mouse cDNA construct was linearized by digestion with EcoRI. The digest was copied with digoxigenin-labeled dUTP by random priming following the Genius 2 DNA labeling kit from Boehringer-Mannheim (Indianapolis, IN, USA) to make probes. The concentration of labeled probe was determined in comparison with labeled standard pBR328 DNA using the dot-blot technique, immunological detection with anti-digoxigenin alkaline phosphatase conjugate (anti-DIG-AP) and the colorimetric method (Boehringer-Mannheim).

Samples of nuclear DNA from skate and four bony fish species (*Esox niger*, *Puntius gonionatus*, *Anguilla rostrata*, and *Acipenser fulvescens*) were denatured by boiling. A 1 µg portion of DNA from each sample was blotted in separate dots on a nylon membrane strip, dried at 80°C for 2 h, and hybridized with 10 ng/ml heat denatured digoxigenin-labeled OC probe in the presence of 50% formamide at 42°C for more than 12 h. The hybrids were detected by a colorimetric method: the hybrid membrane was exposed to 15 U/100 ml anti-DIG-AP conjugate immunodetection buffer for 30 min at room temperature followed by exposure to 0.1 ml/cm^2 color substance containing 0.34 mg/ml nitroblue tetrazolium and 0.175 mg/ml-5-bromo-4-chloro- 3-indolyl phosphate.

Results

Shark and skate skin both appeared to contain very small amounts of OC. The gelatinous nature of the skin extracts made it extremely difficult to purify the protein in greater yield. However, there was a small peak, approximately 10 µg, at the position where OC is normally eluted (Fig. 1). Confirmation of this identity is in progress.

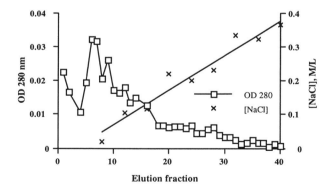

Fig. 1 Elution pattern of skate skin extract from Bio-Rad EconoQ column. The small peak at fraction 28 contained about 5 µg of protein, confirmed by BCA reaction.

Molecular techniques yielded further confirmation, as shown in Fig. 2, of the presence of the OC gene in genomic DNA of the skate as well as in DNA from four species of bony fish.

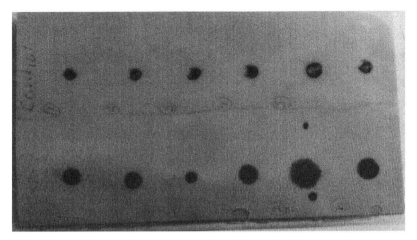

Fig. 2 Dot-blot hybridization of digoxigenin-labeled mouse cDNA with genomic DNA from fish (bottom row) and unlabeled mouse cDNA as control (top row). Fish, left to right are *Raja* male, *Raja* female, *Esox*, *Puntius*, *Anguilla* and *Acipenser* species as noted in the text.

Discussion

These experiments consitute preliminary findings which allow us to determine the location and genetic similarity of OC in a variety of fish including elasmobranchs and more primitive fish, which do not have mineralized bone. Cross-hybridization experiments show that the skate gene shares a common sequence with the other fish species and the mouse cDNA. Interestingly, the presence of the OC gene in *Acipenser*, a primitive ray-finned fish, suggests that OC may have evolved before the development of mineralized endoskeletal bone. Further experiments are necessary to determine the extent of sequence conservation among the various fish OC genes.

Protein comparisons indicate that 28 of the 45-47 amino acids (60%) in the three published bony fishes OC sequences are homologous, while 16 of the fish residues (35%) are homologous with those of mouse OC. The ready hybridization of mouse cDNA with five different fish DNA preparations suggests that OC may be a useful tool in tracing the evolutionary relationships among a wide range of vertebrates. Development of antibodies to purified OC from fish species of interest would enable experiments similar to those using rats to confirm that pOC is an important factor in the stress response.

Acknowledgements
The authors thank D Hinchliffe and K Purohit for their assistance with these experiments.

References

Cancela ML, Williamson MK & Price PA (1995) Amino-acid sequence of bone Gla protein from the African clawed toad *Xenopus laevis* and the fish *Sparus aurata*. *International Journal of Peptide and Protein Research* **46** 419-423.

Delmas PD, Malaval L, Arlot ME & Meunier PJ (1985) Serum bone gla-protein compared to bone histomorphometry in endocrine diseases. *Bone* **6** 339-341.

Ducy P, Desbois D, Boyce B, Pinero G, Story B, Dunstan C, Smith E, Bonadio J, Goldstein S, Gundberg C, Bradley A & Karsenty G (1996) Increased bone formation in osteocalcin-deficient mice. *Nature* **382** 448-452.

Glowacki J & Lian JB (1987) Impaired recruitment and differentiation of osteoclast progenitors by osteocalcin-deplete bone implants. *Cell Differentiation* **21** 247-254.

Gundberg CM, Grant FD, Conlin PR, Chen CJ, Brown EM, Johnson PJ & LeBoff MS (1991) Acute changes in serum osteocalcin during induced hypocalcemia in humans. *Journal of Clinical Endocrinology and Metabolism* **72** 438-443.

Hauschka, PV, Lian BJ, Cole DE & Gundberg CM (1989) Osteocalcin and matrix Gla protein: vitamin K-dependent proteins in bone. *Physiological Reviews* **69** 990-1047.

Hotchkiss CE, Brommage R, Du M & Jerome CP (1998) The anesthetic isoflurane decreaes ionized calcium and increases parathyroid hormone and osteocalcin in cynomolgus monkeys. *Bone* **23** 479-484.

Kvetnansky R & Mikulaj L (1970) Adrenal and urinary catecholamines in rats during adaptation to repeated immobilization stress. *Endocrinology* **87** 738-743.

Napal J, Amado JA, Riancho JA, Olmos JM & Gonzalez-Macias J (1993) Stress decreases the serum level of osteocalcin. *Bone and Mineral* **21** 113-118.

Nishimoto SK (1990) A colorimetric assay specific for gamma-carboxyglutamic acid-containing proteins: its utility in protein purification procedures. *Analytical Biochemistry* **186** 273-279.

Nishimoto SK, Araki N, Robinson FD & Waite JH (1992) Discovery of bone gamma-carboxyglutamic acid protein in mineralized scales. The abundance and structure of *Lepomis macrochirus* bone gamma-carboxyglutamic acid protein. *Journal of Biological Chemistry* **267** 11600-11605.

Patterson-Buckendahl PE, Grindeland RE, Shakes DC, Morey-Holton ER & Cann CE (1988) Circulating osteocalcin in rats is inversely responsive to changes in corticosterone. *American Journal of Physiology* **254** R828-R835.

Patterson-Buckendahl P, Kvetnansky R, Fukuhara K, Cizza G, Kopin I & Cann C (1995) Effects of environmental stress on serum osteocalcin: interaction of corticosterone and norepinephrine. *Bone* **17** 467-472.

Patterson-Buckendahl PE, Cann CE & Kvetnansky R (1996) Osteocalcin response to stress: is it a stress hormone? *Stress: Molecular, Genetic and Neurobiological Advances*, pp 579-589. Eds RN McCarty, G Aguilera, E Sabban & R Kvetnansky. New York: Gordon and Breach.

Patterson-Buckendahl P, Poppalardo D, Kvetnansky R, Globus R, Bikle D, Halloran B, and Morey-Holton E (1996) Opposing effects of vitamin D and stress hormones on bone osteocalcin concentration. *Journal of Bone and Mineral Research* **11** S425 (Abstract).

Price PA, Otsuka AA, Poser JW, Kristaponis J & Raman N (1976) Characterization of a gamma-carboxyglutamic acid-containing protein from bone. *Proceedings of the National Academy of Sciences of the USA* **73** 1447-1451.

Rice JS, Williamson MK & Price PA (1994) Isolation and sequence of the vitamin K-dependent matrix Gla protein from the calcified cartilage of the soupfin shark. *Journal of Bone and Mineral Research* **9** 567-76.

Calcium Metabolism: Comparative Endocrinology
Eds J Danks, C Dacke, G Flik & C Gay, pp 45-48
BioScientifica Ltd, Bristol (1999)

The calcium-sensing receptor in fishes

P M Ingleton[1], P A Hubbard[2], J A Danks[3], G Elgar[4],
R A Sandford[4] and R J Balment[2]

[1]Institute of Endocrinology, Medical School, Sheffield S10 2RX, UK, [2]Department of Biological Sciences, Manchester University, Manchester M13 9PT, UK, [3]St Vincent's Institute of Medical Research, Victoria 3065, Australia and [4]UK-HGMP, Wellcome Trust Campus, Hinxton, Cambridge CB10 1SB, UK

Introduction

For control of normal physiological processes it is essential that calcium ion concentrations are maintained within strict limits both inside the cell and in the circulation. In tetrapod vertebrates, extracellular calcium concentrations are controlled principally by three factors, calcitonin, vitamin D_3 and parathyroid hormone. An essential factor in the control mechanisms for mammalian calcium homoeostasis is the calcium ion itself, which is detected by membrane-located specific receptors in sensitive tissues, particularly the parathyroid gland, brain and kidney (Brown *et al.* 1994, Chattopadyhay *et al.* 1996). The calcium-sensing receptor is a seven-helix-transmembrane domain G-protein-linked receptor which allows extracellular calcium to act as a first messenger. The system is different in fishes because they do not have a parathyroid gland, but control plasma calcium via gills, kidney and intestine, principally (Flik *et al.* 1995).

Fish differ from terrestrial vertebrates in being surrounded by water containing dissolved ions, which may be of varying concentrations, so that, unlike tetrapods, they are not solely dependent upon food as the source of calcium. They therefore need systems for sensing ambient ion concentrations as well as mechanisms for detecting internal circulating ions. In the absence of a parathyroid gland it is likely that several tissues will be involved in control of calcium ion physiology as discussed by Flik *et al.* (1995), but so far an endogenous method of detecting calcium ion concentrations has not been demonstrated in fish. We have now shown that fish tissues express a calcium-sensing receptor (CaSR) gene and contain immunoreactive receptor protein.

Materials and methods

The CaSR has been cloned in the puffer fish (*Fugu rubripes*) and we have prepared an antiserum to an oligopeptide of the extracellular region of the receptor for use in immunohistochemical studies of *Fugu* and flounder (*Platichthys flesus*) tissues. An oligonucleotide sequence was selected from the extracellular domain of the receptor and used as a probe for *in situ* hybridisation studies of the same tissue samples. The *Fugu* tissues were used to determine the normal distribution of the receptor protein in the species of origin of the immunogen. *Fugu* were maintained in aerated seawater at 21±1°C. Tissues were collected from fish anaesthetised in phenoxyethanol and killed by decapitation. All tissues were fixed in sublimated Bouin-Hollande (Kraicer *et al.* 1967) and wax embedded. Sections were mounted on APES-coated slides, and protein and gene expression detected by immunohistochemistry and *in situ* hybridisation, essentially as described in Danks *et al.* (1993, 1997).

Results

CaSRs were found to be widely distributed in *Fugu* tissues; the results of some of the reactions are illustrated in Fig. 1. Epithelial cells of tissues associated with calcium ion control in fish such as the epidermis and kidney tubules possessed CaSRs as well as the crypt cells of the gills ('chloride' cells). Smooth muscle of the gill, skin, gut, gall bladder and air bladder all contained CaSRs. However, smooth muscle of the urinary bladder did not have CaSRs whereas the mainly single layered cells of the luminal epithelium had receptors throughout the cytoplasm, with a concentration at the apical pole. The epithelial cells of the wall of the air bladder contained little CaSR protein but the underlying muscle contained much more. The pattern of protein expression is consistent with the activity of tissues shown to be involved in calcium ion balance and also in those that are dependent upon stable extracellular calcium concentrations for their physiological actions, such as smooth muscle cells lining bladder epithelia, which need to constantly expand and contract.

Another system in which calcium ion concentrations must be controlled is nervous tissue, as axonal neurotransmission depends on regulation of intra- and extra-cellular calcium ions. CaSR protein was found extensively throughout the neuropile of *Fugu* brain and spinal cord. Axons of the pituitary neurohypophysis contained CaSR protein, as did some large neurones of the mid-brain, whereas giant neurones in the brain stem and the anterior spinal cord were mainly negative. Ependymal cells lining the central canal of the spinal cord and choroid plexus epithelium of the mid-brain all possessed CaSRs. Epithelial cells of the saccus vasculosus and their coronets as well as nerves within the organ all contained CaSRs.Within the pituitary, as well as axons of the neurophypophysis, large cells of the intermediate lobe were also immunopositive.The myoendothelial cells of the brain arteries all contained CaSRs and in the chambers of the heart only the muscle of the thick-walled ventricle contained CaSRs and there were apparently none in the atrial wall.

Investigations of tissues from seawater- and freshwater-adapted flounder revealed a similar pattern of staining, with both muscle and epithelial cells possessing CaSRs. In this species the corpuscles of Stannius were collected and examined. The corpuscles, a secretory organ restricted to actinopterygian fishes, produce hypocalcaemic agents, principally stanniocalcins (Butkus *et al.* 1987, Verbost *et al.* 1993) and teleocalcin (Wagner *et al.* 1986). Immunostaining of the flounder corpuscles of Stannius with antiserum to *Fugu* CaSR showed that almost all the cells contained receptor protein, although in seawater fish there was a predominant location in the apical membranes whereas in freshwater-adapted flounder the receptor was distributed

Calcium Metabolism: Comparative Endocrinology

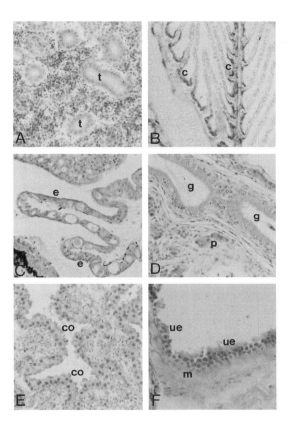

Fig. 1 Calcium-sensing receptors in some tissues of the marine teleost, *Fugu rubripes*. (A) Immunohistochemistry with antiserum to an oligopeptide of the external domain of Fugu calcium-sensing receptor (IHC) of kidney tissue, showing reaction throughout the cytoplasm of kidney tubule (**t**) epithelial cells. (B) In *situ* hybridisation of calcium sensing receptor mRNA in Fugu gill. Hybridisation is most abundant in the crypt or chloride cells (**c**) of the gill lamellae. (C) IHC of the receptor in the epithelial cells (**e**) of the epidermis. (D) ICH of the gut (**g**) and pancreas (**p**); the receptor is most abundant on the luminal border of the gut epithelial cells. (E) IHC of the saccus vasculosus showing the receptor in the cytoplasm of the coronet cells (**co**). (F) Wall of the urinary bladder showing receptors in both the epithelial cells and the underlying smooth muscle (**m**). Bladder epithelium (**ue**).

more evenly in the cells, with concentrations at the basal pole in some cells. Receptors were present in the flounder brain, with the distribution in seawater-adapted fish similar to that in the *Fugu*, also a marine species. However, in freshwater-adapted flounder the receptor protein was depleted from the brain neuropile and from ependymal cells of the posterior recesses of the third ventricle which form nuclei (nucleus recessus posterioris and lateralis) supplying neurotransmitters via axons to the pituitary.

Discussion

Cells in several tissues of both *Fugu* and flounder have calcium-sensing receptors which are distributed throughout the cytoplasm or concentrated at one pole of the cell. The tissues in which the receptor was detected included those concerned with homoeostasis of circulating calcium, for example gut epithelium, and those that depend upon calcium for their normal function, such as nerve and muscle cells. Ambient salinity also appeared to affect calcium-sensing receptors in neural tissue of the flounder. Further physiological studies are needed to characterise the role of the receptor in the different tissues and to define the control of its expression and function.

References

Brown EM, Pollak M & Herbert SC (1994) Cloning and characterisation of extracellular Ca^{2+}-sensing receptors from parathyroid and kidney: molecular physiology and pathophysiology of Ca^{2+}-sensing. *Endocrinologist* **4** 419-426.

Butkus A, Roche PJ, Fernley RT, Haralambidis J, Penschow JD, Ryan GB, Tahair JF, Tregear GW & Coghlan JP (1987) Purification and cloning of a corpuscles of Stannius protein from *Anguilla australis*. *Molecular and Cellular Endocrinology* **54** 123-133.

Chattopadhyay N, Mithal A & Brown EM (1996) The calcium-sensing receptor: window into the physiology and pathophysiology of mineral ion metabolism. *Endocrine Reviews* **17** 289-307.

Danks JA, Devlin AJ, Ho, PMW, Diefenbach-Jagger H, Power DM, Canario A, Martin TJ & Ingleton PM (1993) Parathyroid hormone-related protein is a factor in normal fish pituitary. *General and Comparative Endocrinology* **92** 201-212.

Danks JA McHale JC, Martin TJ & Ingleton PM (1997) Parathyroid hormone-related protein in tissues of the emerging frog (*Rana temporaria*): immunohistochemistry and *in situ* hybridisation. *Journal of Anatomy* **190** 239-260.

Flik G, Verbost PM & Wenderlaar Bonga SE (1995) Calcium transport processes in fishes. In *Cellular and Molecular Approaches To Fish Ionic Regulation*, pp 317-342. New York: Academic Press Inc.

Kraicer J, Herlant M & Duclos P (1967) Changes in adenohypophyseal cytology and nucleic acid content in the rat 32 days after bilateral adrenalectomy and the chronic injection of cortisol. *Canadian Journal of Pharmacology and Physiology* **45** 947-956.

Verbost PM, Butkus A, Willems P & Wendelaar Bonga SE (1993) Indication for two bioactive principles in the corpuscles of Stannius. *Journal of Experimental Biology* **177** 243-252.

Wagner GF, Hamphong M, Park CM & Copp DH (1986) Purification, characterisation and bioassay of teleocalcin, a glycoprotein from salmon corpuscles of Stannius. *General and Comparative Endocrinology* **63** 481-491.

Parathyroid hormone-related protein: localisation in cartilaginous fish tissues

M K Trivett[1,2], T I Walker[3], J G Clement[4], P M Ingleton[5], T J Martin[1] and J A Danks[1]

[1]St Vincent's Institute of Medical Research, 41 Victoria Parade, Fitzroy, 3065 Victoria, Australia, [2]Department of Zoology, University of Melbourne, Parkville 3052, Victoria, Australia, [3]Marine and Freshwater Resources Institute, P.O. Box 114, Queenscliff, 3225 Victoria, Australia, [4]School of Dental Science, University of Melbourne, Parkville, 3052 Victoria, Australia, [5]Institutes of Endocrinology and Cancer Studies, Medical School, Sheffield S10 2RX, UK

Introduction

Parathyroid hormone-related protein (PTHrP) was discovered through its role in the elevation of plasma calcium levels associated with some types of tumours (humoral hypercalcaemia of malignancy) (Moseley *et al.* 1987, Suva *et al.* 1987). It seems likely that PTH and PTHrP evolved from a common ancestral gene (Mangin *et al.* 1988, Yasuda *et al.* 1989). While PTH is a classical hormone in land-dwelling vertebrates, it appears that PTHrP may only act as a hormone during pregnancy and lactation in mammals (review: Martin *et al.* 1997). PTHrP is widely distributed in mammalian (see Martin *et al.* 1997) and avian (Schermer *et al.* 1991) tissues. It appears to have a number of physiological roles in mammals, including the modulation of growth and differentiation, foetal development, transport of calcium across the placenta and smooth muscle relaxation, all of which seem to be mediated by the paracrine or autocrine actions of PTHrP (Martin *et al.* 1997).

Lower vertebrates also appear to possess PTHrP. An immunohistochemical study using antibodies to N-terminal human PTHrP demonstrated immunoreactive PTHrP in the frog *Rana temporaria* (Danks *et al.* 1997). Immunoreactive PTHrP has also been found in the pituitary of the sea bream *Sparus aurata* (Danks *et al.* 1993) and in the pituitary, kidney and rectal gland of the dogfish *Scyliorhinus canicula* (Ingleton *et al.* 1995). These studies suggest that PTHrP is widely distributed in lower vertebrates, but only a limited number of species and tissues have been examined. The presence of immunoreactive PTHrP has not been investigated in lampreys or hagfish, the living representatives of the oldest group of vertebrates, the agnathans. PTHrP-like DNA has been detected in hagfish (Chailleux *et al.* 1995), but the cellular distribution of PTHrP mRNA in fish has not been examined.

This study examined the cellular distribution of immunoreactive PTHrP and PTHrP mRNA in the lamprey *Geotria australis* and in a range of elasmobranch species. These results may be used to speculate on the roles of PTHrP in fish and to examine the evolutionary history of this molecule.

Materials and methods

Tissues

All tissues were formalin fixed and paraffin embedded. Two whole downstream migrant (ds) adults and four larvae (ammocoetes) of the lamprey *G. australis* were generously provided by Professor Ian Potter and Glen Power (University of Western Australia, Western Australia, Australia).

Kidney, gill, rectal gland, skin and pituitary samples were taken from gummy sharks (*Mustelus antarcticus*), school sharks (*Galeorhinus galeus*), angel sharks (*Squatina australis*), Port Jackson sharks (*Heterodontus portusjacksoni*), common spotted stingarees (*Urolophus gigas*) and banjo sharks (*Trygonorrhina fasciata*). Samples of vertebrae and jaw were taken from gummy and school sharks.

Immunohistochemistry

Rabbit antisera, raised to synthetic human N-terminal PTHrP(1-14) and (1-16) were used for immunohistochemistry. These antisera showed no cross-reactivity with human parathyroid hormone, and the antiserum to PTHrP(1-14) has been successfully used on fish tissues (Danks *et al.* 1993, Ingleton *et al.* 1995). PTHrP immunohistochemistry followed the protocol previously described by Danks *et al.* (1993) and Ingleton *et al.* (1995).

Controls

Human skin served as a positive tissue control in each experiment. Non-immune rabbit serum substituted for the primary antibody served as the negative control. The deletion of alternate layers of the antibody sandwich served as a method of control.

Sections were stained with antiserum to human PTH(1-34) (BioGenex) as previously described (Danks *et al.* 1993) to confirm the specificity of the PTHrP staining.

In situ hybridisation

The probe used in these studies was against exon VI of the human PTHrP gene, a region that shows conservation among mammalian species. The probe was a 420 bp riboprobe labelled with digoxygenin and has been used to examine PTHrP expression in mammalian tissues (Kartsogiannis *et al.* 1997). The protocol used was that of Zhou *et al.* (1994) and Kartsogiannis *et al.* (1997).

Controls

Human skin served as the positive control tissue in each experiment, while sections treated with 100 µg/mL RNase, before hybridisation, acted as a negative control. Antibody specificity was confirmed using no-probe controls with antibody, and endogenous alkaline phosphatase activity was assessed using no-probe slides with antibody omitted as a standard. RNase and no-probe controls were included for every tissue within each experiment. Hybridisation was deemed non-specific if the signal was inconsistent between and within samples or if it did not localise specifically with cells in the tissue.

Results

The patterns of staining for PTHrP antigen and mRNA were similar in all the elasmobranchs studied. PTHrP antigen and mRNA were detected in the elasmobranchs and in the lamprey in osmoregulatory tissues such as the gill, kidney, rectal gland and pituitary (Table 1 and Fig. 1). The pattern of PTHrP mRNA distribution generally reflected that of the immunoreactive PTHrP.

Table 1 Osmoregulatory tissues.

Tissue	Method	PTHrP distribution
Gill		
Sharks, rays, ds migrant	IHC	• Strongest in primary gill epithelium • Weak in secondary lamellar epithelium • Some pillar cells contained PTHrP • PTHrP not seen in connective tissue
Larva	IHC	• Diffuse staining in primary and secondary lamellae
All species	ISH	• Strong signal in primary lamellar epithelium • Weak signal in secondary lamellae
Kidney		
All species	IHC	• Staining in proximal and distal tubules • Weak staining in glomeruli • No staining in neck segments
All species	ISH	• Signal in glomeruli, neck segments, proximal and distal tubules
Rectal gland		
Sharks and rays	IHC and ISH	•Unique to sharks and rays •Tubular and lumen epithelium positive •Connective tissue negative
Pituitary		
Sharks and rays	IHC and ISH	•Not studied in lampreys •Staining in endocrine cells of pars distalis and pars intermedia

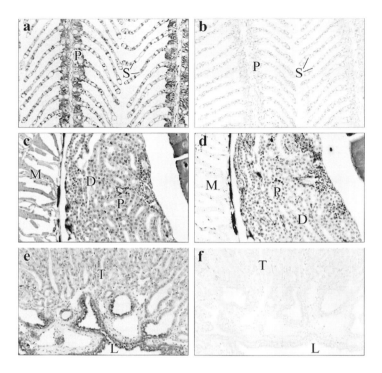

Fig. 1 PTHrP in osmoregulatory tissues. (a) *In situ* hybridisation with PTHrP riboprobe in shark gill. Signal for the PTHrP mRNA is strongest in the primary lamellar epithelium (P) but weaker in the secondary lamellar epithelium (S). Signal is also seen in some of the pillar cells that divide the capillary channel of the secondary lamellae. (b) Section of shark gill treated with RNase before hybridisation. Only the counterstain is visible; note the absence of signal from the primary (P) and secondary (S) lamellar epithelium. (c) Immunohistochemistry of lamprey kidney. Moderate PTHrP antigen staining is seen in the proximal (P) and distal tubules (D) of the kidney. Stronger staining is seen in the nearby skeletal muscle (M). A layer of pigment is seen on the inner edge of the muscle. (d) Non-immune control of lamprey kidney. Only the haematoxylin counter stain is visible in the proximal (P) and distal tubules (D) and the muscle (M). The layer of dark pigment on the inner edge of the muscle is also visible. (e) *In situ* hybridisation of shark rectal gland. Signal for PTHrP mRNA is seen in the epithelial cells of the tubules (T) and in epithelium surrounding the central lumen (L). (f) RNase control of shark rectal gland. Signal is absent from the tubular epithelium (T) and the epithelium around the central lumen (L).

Immunoreactive PTHrP and PTHrP mRNA were demonstrated in skeletal tissues in both lamprey and elasmobranchs (Table 2 and Fig. 2). A similar distribution was noted for the immunoreactive PTHrP and PTHrP mRNA.

PTH immunoreactivity was not observed in any of the tissues examined in this study (data not shown).

Discussion

The lampreys used in this study were resident in fresh water, whereas the elasmobranchs studied were marine. PTHrP distribution in the kidney of both these groups was similar to that described for mammalian species (Soifer et al. 1993, Yang et al. 1997). PTHrP has been implicated in the modulation of cellular growth and differentiation in a variety of tissues (review: Martin et al. 1997), including renal epithelium (Garcia-Ocana et al. 1995), and it is possible that it has a similar role in fish renal epithelium. In addition, PTHrP appears to affect the renal handling of calcium and phosphate in the mammalian kidney (Ebeling et al. 1989, Zhou et al. 1989) and may have similar actions on the fish kidney.

Immunoreactive PTHrP was not detected in dogfish gill (Ingleton et al. 1995), but its distribution in the downstream migrant lamprey gill and the elasmobranch gill is similar to that recently described for flounder (Danks et al. 1998). The lamprey ammocoete undergoes radical metamorphosis to become an adult and the differences in the distribution of immunoreactive PTHrP in the ammocoete gill and the downstream migrant gill may be related to differences in the stage of development and structure of the gill in these two phases of the life cycle. It is possible that PTHrP modulates the growth and differentiation of the epithelium in the fish gill, as has been shown for a variety of epithelial cell types in mammals (Martin et al. 1997).

The elasmobranch rectal gland has roles in the regulation of sodium and chloride (Haywood 1975). Immunohistochemistry detected PTHrP in the elasmobranch rectal gland (Ingleton et al. 1995), and the hybridisation results from the present study suggest that this immunoreactive PTHrP is probably produced *in situ* by the tubule cells. It is possible that PTHrP may have roles in the growth and differentiation of the epithelial cells of the tubules, as it does in mammalian epithelial cells (Martin et al. 1997). The possible roles of PTHrP in the regulation of sodium and chloride transport by the rectal gland need to be investigated physiologically.

Immunohistochemistry demonstrated PTHrP antigen in the dogfish pituitary (Ingleton et al. 1995), and results from the present study demonstrate that PTHrP occurs in the pituitary of distantly related elasmobranchs. The presence of PTHrP mRNA in the same sites as PTHrP antigen suggests that the elasmobranch pituitary produces PTHrP *in situ*. The teleost pituitary releases PTHrP (Danks et al. 1993), and studies with pituitary cultures may demonstrate a similar finding in the elasmobranch pituitary.

PTHrP antigen and mRNA were consistently demonstrated in skeletal tissues from elasmobranchs and the lamprey. The notochord forms part of the axial skeleton in all vertebrate embryos. It becomes constricted by the vertebral column in adult elasmobranchs, but persists throughout life in the lampreys. The presence of PTHrP in

Table 2 Skeletal tissues.

Tissue	Method	PTHrP distribution
Notochord		
Lamprey*	IHC and ISH	• Found in vacuolated notochordal cells and notochordal epithelium • Not seen in notochordal sheath
Shark	IHC and ISH	• Weak in notochordal and epithelial cells • Some positive cells in outer calcified sheath • Inner sheath negative
Vertebrae		
Lamprey*	IHC and ISH	• Positive chondrocytes in neural arches • Cells of perichondrium positive • PTHrP not in connective tissue
Shark	IHC and ISH	• Positive chondrocytes in neural arch • Chondrocytes deep within cartilage were negative • Chondrocytes at calcification front contained PTHrP • Chondrocytes in calcified cartilage did not contain PTHrP • Cells in perichondrium contained PTHrP
Jaw		
Shark	IHC and ISH	• PTHrP seen in perichondrium • PTHrP not observed deeply embedded chondrocytes • PTHrP not present in tesserae (calcified cartilage)
Teeth and developing denticles		
Shark	IHC	• PTHrP observed in ameloblasts and odontoblasts (epithelium) • Scattered cells in the pulp cavity stained
	ISH	• Signal in all epithelial layers • Signal within pulp cavity
Branchial and fin cartilage		
Lamprey	IHC and ISH	• Chondrocytes within areas of proliferation contained PTHrP

* Pattern of staining was the same in the ammocoete and ds migrant.

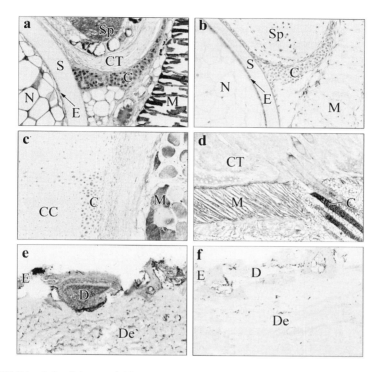

Fig. 2 PTHrP in skeletal tissues. (a) Immunohistochemistry of lamprey vertebrae. PTHrP antigen is seen in the notochordal cells (N), in the notochordal epithelium (E), skeletal muscle (M) and in chondrocytes of the neural arch (C). PTHrP was not seen in connective tissue (CT) or in the sheath (S) surrounding the notochord. Part of the spinal cord (Sp) is visible. (b) Non-immune control for (a). Only staining by haematoxylin is visible in the chondrocytes (C), notochordal cells (N), notochordal epithelium (E) and sheath (S). Part of the spinal cord (Sp) is visible. (c) Immunohistochemistry of shark vertebrae. Chondrocytes (C) in the uncalcified matrix contained PTHrP while those in the calcified cartilage (CC) did not. PTHrP antigen was also seen in skeletal muscle (M). (d) *In situ* hybridisation of lamprey fin. PTHrP mRNA was seen in chondrocytes (C) in the cartilaginous fin supports and skeletal muscle (M). Signal was not seen in the connective tissue (CT). (e) *In situ* hybridisation of shark skin. Specific signal of PTHrP mRNA is seen in the epithelial layers of the developing denticle (D) and in the epidermis (E), but not in the underlying dermis (De). (f) RNase control of shark skin showing an absence of signal in the developing denticle (D), epidermis (E) and dermis (De).

the notochord suggests that PTHrP may have an ancient association with skeletal tissues other than cartilage and bone. In addition, the notochord pre-dates vertebrates and is found in all chordates (Hildebrand 1988). It is possible that examination of the invertebrate chordate notochord will reveal that PTHrP is present in animals that precede vertebrate evolution.

The elasmobranchs and lampreys both possess a cartilaginous endoskeleton. The presence of PTHrP in the chondrocytes of these early vertebrates suggests that PTHrP may have an ancient association with vertebrate cartilage. It may have similar roles in fish chondrocytes to those in mammalian chondrocytes, such as modulation of proliferation and differentiation (Martin *et al.* 1997). Unlike the lamprey skeleton, the shark skeleton contains pockets of calcified cartilage, and formation of this skeleton is similar to endochondral bone formation in mammals (Clement 1992). PTHrP has been detected in proliferating, pre-hypertrophic and hypertrophic chondrocytes during mammalian endochondral bone formation, although the intensity of expression changes during development (Lee *et al.* 1995, Kartsogiannis *et al.* 1997). A study of the skeleton in foetal sharks will be necessary to discover how closely the distribution of PTHrP in a calcified cartilaginous skeleton resembles that in endochondral bone formation. It may be that PTHrP is conservatively localised to a tissue type, rather than a specific cell type, during skeletal development in vertebrates.

The teeth and dermal denticles in elasmobranchs undergo constant replacement throughout the life of the animal. PTHrP was found in elasmobranch teeth and dermal denticles in similar sites, consistent with the idea that these structures form by the same process (Kendall 1947). In addition, the formation of shark teeth (and denticles) follows a similar process to mammalian tooth development (Moss 1977). PTHrP has previously been detected in the epithelia of developing rat teeth (Beck *et al.* 1995), and the similarity between PTHrP distribution in developing shark and rat teeth suggests that PTHrP may have similar roles in shark tooth development to those suggested in the rat, perhaps in cell division.

This study has shown that PTHrP mRNA and protein are widely distributed in distantly related fish, and the sites of distribution suggest that PTHrP may have roles in osmoregulatory and skeletal tissues in fish. The presence of PTHrP in the lamprey suggests that PTHrP is of ancient origin, and it is possible that it will be found in invertebrate chordates. The successful use of antibodies and a probe to human PTHrP sequence suggests that the PTHrP molecule in fish may be similar to that in humans. The conservation of the PTHrP in distantly related vertebrates suggests that it may have fundamental and basic physiological roles in all vertebrates.

Acknowledgements

We thank David Paul (Department of Zoology, Univeristy of Melbourne) for his help with the photomicrography and Dr Hong Zhou (Department of Medicine, University of Melbourne) for preparing the PTHrP riboprobe. We thank researchers at the Marine and Freshwater Resources Institute for obtaining the elasmobranchs.

References

Beck F, Tucci J, Russell A, Senior P & Ferguson M (1995) The expression of the gene encoding parathyroid hormone-related protein (PTHrP) during tooth development in the rat. *Cell and Tissue Research* **280** 283-290.

Chailleux N, Milet C, Vidal A & Lopez E (1995) Presence of PTH-like and PTH-related peptide-like molecules in submammalian vertebrates. *Netherlands Journal of Zoology* **45** 248-250.

Clement J (1992) Re-examination of the fine structure of endoskeletal mineralization in chondrichthyans: implications for growth, ageing and calcium homeostasis. *Australian Journal of Marine and Freshwater Research* **43** 157-181.

Danks J, Devlin A, Ho P, Diefenbach-Jagger H, Power D, Canario A *et al.* (1993) Parathyroid hormone-related protein is a factor in normal fish pituitary. *General and Comparative Endocrinology* **92** 201-212.

Danks J, McHale J, Martin T & Ingleton P (1997) Parathyroid hormone-related protein in tissues of the emerging frog (*Rana temporaria*): immunohistochemistry and *in situ* hybridisation. *Journal of Anatomy* **190** 229-238.

Danks J, Hubbard P, Balment R, Ingleton P & Martin T (1998) Parathyroid hormone-related protein localization in tissues of freshwater and saltwater-acclimatized flounder. *Annals of the New York Academy of Science* **839** 503-505.

Ebeling P, Adam W, Moseley J & Martin T (1989) Actions of synthetic parathyroid hormone-related protein (1-34) on the isolated rat kidney. *Journal of Endocrinology* **120** 45-50.

Garcia-Ocana A, De Miguel F, Penaranda C, Albar J, Sarasa J & Esbrit P (1995) Parathyroid hormone-related protein is an autocrine modulator of rabbit proximal tubule cell growth. *Journal of Bone and Mineral Research* **10** 1875-1884.

Haywood G (1975) A preliminary investigation into the roles played by the rectal gland and kidneys in osmoregulation of the striped dogfish *Poroderma africanum*. *Journal of Experimental Zoology* **193** 167-175.

Hildebrand M (1988) *Analysis of Vertebrate Structure*, New York: John Wiley and Sons.

Ingleton P, Hazon N, Ho P, Martin T & Danks J (1995) Immunodetection of parathyroid hormone-related protein in plasma and tissues of an elasmobranch (*Scyliorhinus canicula*). *General and Comparative Endocrinology* **98** 211-218.

Kartsogiannis V, Moseley J, McKelvie B, Chou S, Hards D, Ng K *et al.* (1997) Temporal expression of PTHrP during endochondral bone formation in mouse and intramembranous bone formation in an *in vivo* rabbit model. *Bone* **21** 385-392.

Kendall J (1947) *Microscopic Anatomy of Vertebrates*, London: Henry Kimpton.

Lee K, Deeds J & Segre G (1995) Expression of parathyroid hormone-related peptide and its receptor messenger ribonucleic acids during fetal development of rats. *Endocrinology* **136** 453-463.

Mangin M, Webb A, Dreyer B, Posillico J, Ikeda K, Weir E *et al.* (1988) Identification of a cDNA encoding a parathyroid hormone-like peptide form a human tumor associated with humoral hypercalcemia of malignancy. *Proceedings of the National Academy of Sciences of the USA* **85** 597-601.

Martin T, Moseley J & Williams E (1997) Parathyroid hormone-related protein: hormone and cytokine. *Journal of Endocrinology* **154** S23-S37.

Moseley J, Kubota M, Diefenbach-Jagger H, Wettenhall R, Kemp B, Suva L *et al.* (1987) Parathyroid hormone-related protein purified from a human lung cancer cell-line. *Proceedings of the National Academy of Sciences of the USA* **84** 5048-5052.

Moss M (1977) Skeletal tissues in sharks. *American Zoologist* **17** 335-342.

Schermer D, Chan S, Bruce R, Nissenson R, Wood W & Strewler G (1991) Chicken parathyroid hormone-related protein and its expression during embryonic development. *Journal of Bone and Mineral Research* **6** 149-155.

Soifer N, Vanwhy S, Ganz M, Kashgarian M, Seigel N & Stewart A (1993) Expression of parathyroid hormone-related protein in the rat glomerulus and tubule during recovery from renal ischemia. *Journal of Clinical Investigation* **92** 2850-2857.

Suva L, Winslow G, Wettenhall R, Hammonds R, Moseley J, Diefenbach-Jagger H *et al.* (1987) A parathyroid hormone-related protein implicated in malignant hypercalcemia: cloning and expression. *Science* **237** 893-896.

Yang T, Hassan S, Huang Y, Smart A, Briggs J & Schnermann J (1997) Expression of PTHrP, PTH/PTHrP receptor, and Ca^{2+}-sensing receptor mRNAs along the rat nephron. *American Journal of Physiology* **41** F751-F758.

Yasuda T, Banville D, Hendy G & Goltzman D (1989) Characterization of the human parathyroid hormone-like peptide gene: functional and evolutionary aspects. *Journal of Biological Chemistry* **264** 7720-7725.

Zhou H, Leaver D, Moseley J, Kemp B, Ebeling P & Martin T (1989) Actions of parathyroid hormone-related protein on the rat kidney *in vivo*. *Journal of Endocrinology* **122** 229-235.

Zhou H, Choong P, McCarthy R, Chou S, Martin T & Ng K (1994) *In situ* hybridization to show sequential expression of osteoblast gene markers during endochondral bone formation *in vivo*. *Journal of Bone and Mineral Research* **9** 1489-1499.

Molecular cloning of cDNAs encoding three distinct receptors for parathyroid hormone (PTH)/PTH-related peptide in the zebrafish

D A Rubin and H Jüppner

Endocrine Unit, Department of Medicine, Massachusetts General Hospital and Harvard Medical School, Boston, Massachusetts 02114, USA

Introduction

Parathyroid hormone (PTH)- and PTH-related peptide (PTHrP)-like immunoreactivity has long been described in several fishes (Harvey et al. 1987, Kaneko & Pang 1987, Ingleton et al. 1995, Devlin et al. 1996). Indeed, partial DNA sequences encoding a putative fish PTH molecule were isolated from Rainbow trout genomic DNA (Rosenberg & Kronenberg 1991), and a genomic DNA sequence possibly encoding a teleost PTHrP homolog was recently isolated from pufferfish (FUGU Landmark Mapping Project Database; available from http://www.fugu.hgmp.mrc.ac.uk). While the major biological function of PTH is well defined in mammals, functions mediated by PTH-like or PTHrP-like ligands in fishes remain largely elusive. To begin exploring the evolution of these peptides, we searched for cDNAs encoding their receptors, because these proteins are likely to show higher amino acid conservation than their cognate ligands. Furthermore, members of this family of G-protein-coupled receptors have already been found in insects, nematodes, and goldfish (Reagan 1994, Chan et al. 1998). Our efforts led to the isolation of cDNAs encoding three distinct zebrafish receptors that are activated by mammalian PTH and/or PTHrP. These findings suggest that PTH-like or PTHrP-like molecules exist in non-tetrapod vertebrates, the teleosts.

Cloning and function of the zebrafish homolog of the mammalian PTH type-2 receptor (PTH2R)

The first genomic DNA sequence to indicate that PTH/PTHrP-like receptors (Jüppner et al. 1991, Kong et al. 1994, Schipani et al. 1995) may be present in fishes was reported by Hellman & Jüppner (1993) for the catfish (*Ictalurus punctatus*). This partial genomic sequence was subsequently correctly identified as being homologous to the mammalian PTH2R (Usdin et al. 1995). With the available genomic DNA sequence, reverse transcriptase PCR was performed on catfish liver and kidney total RNA to generate a cDNA that encoded the equivalent of transmembrane 3 through 5 of the mammalian PTH2R (Usdin et al. 1995).

Because catfish and zebrafish are members within the Division Ostariophysi, and since a zebrafish kidney cDNA library was available, catfish-specific primers were used to identify and clone the full-length zebrafish homolog of the mammalian PTH2R

(DA Rubin, P Hellman, LI Zon, CJ Lobb, C Bergwitz and H Jüppner, unpublished observations). The predominant form of this zPTH2R cDNA contains an unusually long sequence downstream of the signal sequence and upstream of the region encoded in mammals by exon E1. 5'-Rapid amplification of cDNA ends (RACE) was performed to isolate a putative splice variant (zPTH2R#43) that was more homologous to the extracellular region of the human PTH2R. This receptor variant showed, when expressed in mammalian COS-7 cells, a significantly higher increase in agonist-induced cAMP accumulation than the predominant form of the zPTH2R, and was therefore characterized further. These functional studies showed that the zPTH2R responds to PTH but not PTHrP, which is similar to the response observed with the human PTH2R (DA Rubin, P Hellman, LI Zon, CJ Lobb, C Bergwitz and H Jüppner, unpublished observations). This conservation of receptor characteristics between two very divergent lineages of vertebrates suggests that a PTH-like peptide (Usdin 1997) is likely to exist in fishes and may serve this species, functions not related to mineral ion homeostasis.

Cloning of partial genomic DNA sequences encoding the zebrafish homolog of the PTH1R and of a novel PTH3R

With the recent isolation of two non-allelic frog PTH1Rs (Bergwitz et al. 1998) and the finding that teleosts express a PTH2R, we designed several degenerate primers based on a region of the PTH1R which is highly conserved in mammals and frogs (Schipani et al. 1995, DA Rubin, P Hellman, LI Zon, CJ Lobb, C Bergwitz and H Jüppner, unpublished observations). PCR using genomic DNA and primers located in exon M6/7 (encoding the third extracellular loop and portions of transmembrane (TM) domains 6 and 7) and exon M7 (encoding portions of TM7) led to the isolation of a DNA sequence that encodes an homologous portion of the tetrapod PTH1R; this clone is refered to as the zPTH1R (DA Rubin and H Jüppner, unpublished observations). Subsequent to the isolation of this partial PTH1R clone, a different genomic DNA product was obtained which was distinct from the zPTH1R and the zPTH2R, but shared significant homology with these two receptors (DA Rubin and H Jüppner, unpublished observations).

Isolation of partial cDNA clones encoding zPTH1R and zPTH3R

Reverse transcriptase PCR on total zebrafish RNA was performed using reverse primers that were based on either the zPTH1R or zPTH3R genomic DNA sequences and a degenerate forward primer in TM3. Two PCR products of about 450 bp were isolated and sequenced. The first product was shown to encode TM3 through TM6 of the zebrafish homolog of the PTH1R (zPTH1R), and the second PCR product was shown to encode a novel receptor, designated PTH3R (zPTH3R) (DA Rubin and H Jüppner, unpublished observations).

Cloning and functional characterization of the zebrafish PTH1R

5'-RACE and 3'-RACE on total zebrafish RNA was performed with primers specific for the zPTH1R sequence to obtain a full-length cDNA, which was cloned into pcDNAI/Amp (Invitrogen, San Diego, California, USA) for transient expression in COS-7 cells and functional characterization. When the zPTH1R was challenged with human (h)PTH or hPTHrP, the two peptides stimulated cAMP accumulation almost equivalently (EC_{50} 0.47±0.33 and 1.11±0.07 nM respectively; means±S.E.M), and both peptides showed high-affinity binding (IC_{50} 3.3±0.90 and 2.2±0.64 nM respectively). Furthermore, hPTH and hPTHrP stimulated total inositol phosphate turnover approximately twofold (DA Rubin and H Jüppner, unpublished observations).

Cloning and function of a novel PTHrP-selective type-3 receptor (PTH3R) from zebrafish.

Using the 450 bp zPTH3R cDNA as a probe, a λgt11 zebrafish library was screened and a single phage (insert size approximately 2.5 kb) was isolated and plaque purified. Because most of the N-terminal extracellular domain was missing, 5'-RACE was performed on total zebrafish RNA, and several putative splice variants were isolated, one of which contained a putative signal peptide with high homology to the zPTH1R

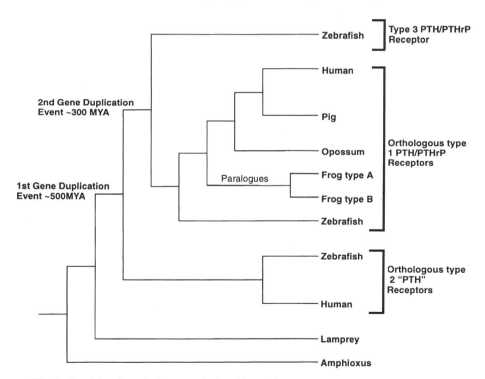

Fig. 1 Provisional evolutionary relationship of known PTH/PTHrP receptors. A previous phylogenetic analysis (DA Rubin and H Jüppner, unpublished observations) generated a single most parsimonious tree. This cladogram is, in part, recreated to indicate the evolutionary history of the three PTH receptor subtypes.

and a Kozak sequence upstream of the initiator AUG. A full-length cDNA encoding the zPTH3R was constructed in pcDNAI/Amp and transiently expressed in COS-7 cells. In contrast with the findings with the zPTH1R, this novel receptor was preferentially activated by hPTHrP (EC_{50} 0.48±0.35 nM) and showed an approximately 20-fold lower efficacy for hPTH. This ligand selectivity was confirmed by radioreceptor studies using ^{125}I-labeled hPTHrP, which showed high-affinity binding of hPTHrP (IC_{50} 2.5±0.98 nM), but much lower binding affinity of hPTH (IC_{50} 91.3±14.5 nM) (DA Rubin and H Jüppner, unpublished observations). Interestingly, no inositol phosphate signaling was observed, which is different from the findings with the zPTH1R (DA Rubin and H Jüppner, unpublished observations).

Southern blot analysis

To confirm that the three zebrafish PTHRs are encoded by distinct genes, genomic DNA was digested to completion with infrequently cutting restriction enzymes, electrophoresed through a 0.8% agarose gel, and hybridized to probes that encode the C-terminal tail of zPTH1R, zPTH2R or zPTH3R, and which are likely to be encoded by a single exon. For each digest these tail probes hybridized, under stringent conditions, to single but distinct genomic DNA fragments indicating that different genes encode the three zebrafish PTH/PTHrP receptors (DA Rubin, P Hellman, LI Zon, CJ Lobb, C Bergwitz and H Jüppner, unpublished observations, DA Rubin and H Jüppner, unpublished observations).

Table 1 Comparisons of the amino acid sequences of the entire coding region from different PTH receptors. Residues comprising the coding region of the zPTH1R and zPTH3R were compared with corresponding regions of the zPTH2R, and the human PTH1R or PTH2R. Values are %Similarity/ %Identity.

	zPTH1R	**zPTH2R**	**zPTH3R**	**hPTH1R**
zPTH2R	61/54	*		
zPTH3R	66/59	55/46	*	
hPTH1R	76/70	53/48	68/59	*
hPTH2R	59/52	69/64	57/49	61/53

Cladistic relationship of PTH/PTHrP receptors

A phylogenetic analysis was performed as previously described ((DA Rubin and H Jüppner, unpublished observations) and modified to include the chordate (amphioxus) and agnatha (lamprey) to evaluate the congruency between the PTHR gene phylogeny and PTHR species phylogeny (Goodman *et al.* 1978). The cladogram indicates that two clades are predominant: the type-2 receptor clade and the PTH1R/PTH3R clade (Fig. 1). Analysis of this modified cladogram for proposed periods of gene duplication within the vertebrates shows it to be in accordance with Ohno's theory of evolution of the vertebrates by gene duplication (Ohno 1970). Furthermore, the phylogenetic relationships of species within the PTH1R clade are in accordance with morphologically based phylogenies (DA Rubin and H Jüppner, unpublished observations).

When the coding region sequences of these zebrafish receptors were compared with human PTH1R and human PTH2R, the type-1 and type-2 receptors shared the highest degree of similarity within each subtype (Table 1). The zPTH3R shared a lower homology with the PTH2R than with the PTH1R, indicating that it tentatively groups within the type-1/type-3 PTH/PTHrP receptor clade. With the cloning of other type-3 receptors from other species, this subtype may become statistically distinguishable from the type-1 receptor.

The tissue-specific expression of the RNAs encoding these different receptor proteins will help to define their biological importance in fishes, and may provide clues for those biological roles in mammals that are likely to be distinct from those required for mineral ion homeostasis. Furthermore, the more detailed characterization of ligand-binding and signaling properties of these non-mammalian proteins will provide opportunities to explore functionally important domains within these receptors and their cognate ligands, and may lead to the identification of agonists, such as salmon calcitonin, that are significantly more potent in humans than the mammalian homologs.

References

Bergwitz C, Klein P, Kohno H, Forman SA, Lee K, Rubin DA & Jüppner H (1998) Identification, functional characterization, and developmental expression of two non-allelic PTH/PTHrP receptor isoforms in *Xenopus laevis* (Daudin). *Endocrinology* **139** 723-732.

Chan KW, Yu KL, Rivier J & Chow BKC (1998) Identification and characterization of a receptor from goldfish specific for a teleost growth hormone-releasing hormone-like peptide. *Neuroendocrinology* **68** 44-56.

Devlin AJ, Danks JA, Faulkner MK, Power DM, Canario AVM, Martin TJ & Ingleton PM (1996) Immunochemical detection of parathyroid hormone-related protein in the saccus vasculosus of a teleost fish. *General and Comparative Endocrinology* **101** 83-90.

Goodman M, Czelusniak J, Moore GW, Romero-Herrera AE & Matsuda G (1978) Fitting the gene lineage into its species lineage, a parsimonious strategy illustrated by cladograms constructed from globin sequences. *Systematic Zoology* **28** 132-163.

Harvey S, Zeng YY & Pang PKT (1987) Parathyroid hormone-like immunoreactivity in fish plasma and tissues. *General and Comparative Endocrinology* **68** 136-146.

Hellman P & Jüppner H (1993) New Proteins in Ca^{2+} regulation: parathyroid hormone-related protein and a Ca^{2+} sensor molecule. Acta Universitas Uppsaliensis University of Uppsala.

Ingleton PM, Hazon N, Ho, PMW, Martin TJ & Danks JA (1995) Immunodetection of parathyroid hormone-related protein in plasma and tissues of an elasmobranch (*Scyliorhinus canicula*). *General and Comparative Endocrinology* **98** 211-218.

Jüppner H, Abou-Samra AB, Freeman M, Kong XF, Schipani E, Richards J, Kolakowski Jr,LF, Hock J, Potts Jr, JT, Kronenber HM & Segre GV (1991) A G protein-linked receptor for parathyroid hormone and parathyroid hormone-related peptide. *Science* **254** 1024-1026.

Kaneko T & Pang PKT (1987) Immunocytochemical detection of parathyroid hormone-like substance in the goldfish brain and pituitary gland. *General and Comparative Endocrinology* **68** 147-152.

Kong XF, Schipani E, Lanske B, Joun H, Karperien M, Defize LHK, Jüppner H, Potts Jr, JT, Segre GV, Kronenberg HM & Abou-Samra AB (1994) The rat, mouse, and human genes encoding the receptor for parathyroid hormone and parathyroid hormone-related peptide are highly homologous. *Biochemical and Biophysical Research Communications* **200** 1290-1299.

Ohno S (1970) *Evolution by Gene Duplication.* Berlin: Springer-Verlag.

Reagan JP (1994) Expression cloning of an insect diuretic hormone receptor: a member of the calcitonin/secretin receptor family. *Journal of Biological Chemistry* **269** 9-12.

Rosenberg J & Kronenberg H (1991) Parathyroid hormone without parathyroid glands: a sequence resembling PTH in a teleost fish. *Journal of Bone and Mineral Research* **6** (Suppl.1) S178.

Schipani E, Weinstein LS, Bergwitz C, Iida-Klein A, Kong XF, Stuhrmann M, Kruse K, Whyte MP, Murray T, Schmidtke J, van Dop C, Brickman AS, Crawford JD, Potts Jr, JT, Kronenberg HM, Abou-Samra AB, Segre GV & Jüppner H (1995) Pseudohypoparathyroidism Type Ib is not caused by mutations in the coding exons of the human parathyroid hormone (PTH)/PTH-related peptide receptor gene. *Journal of Clinical Endocrinology and Metabolism* **80** 1611-1621.

Usdin TB (1997) Evidence for a parathyroid hormone-2 receptor selective ligand in the hypothalamus. *Endocrinology* **138** 831-834.

Usdin TB, Gruber C & Bonner TI (1995) Identification and functional expression of a receptor selectively recognizing parathyriod hormone, the PTH2 receptor. *Journal of Biological Chemistry* **270** 15455-15458.

Part Three

Reptiles and birds

Regulation of the calcium-sensing and parathyroid hormone receptor genes in the chick

M Pines[1], N Yarden[1], S Ben-Bassat[1], I Lavelin[1] and R M Leach[2]

[1]Institute of Animal Science, Agricultural Research Organization, The Volcani Center, Bet Dagan 50250, Israel and [2]Department of Poultry Sciences, Pennsylvania State University, University Park, PA 16802-3501, USA

Introduction

Calcium (Ca) ions are of critical importance for a variety of vital body functions (Brown 1991). Intra- and extra-cellular Ca^{2+} act in distinct but sometimes complementary ways to regulate a multitude of biological processes. In contrast with cytosolic Ca^{2+}, extracellular Ca^{2+} concentration remains almost constant and under normal circumstances varies by only a few percentage points throughout life (Hurwitz 1989, Brown 1991, Kurokawa 1994). A key factor for ensuring mineral ion homeostasis is the presence of Ca^{2+}-sensing cells in the parathyroid gland (PTG); these cells detect minute perturbations in the extracellular Ca^{2+} concentration, and respond with alterations in cellular functions that normalize Ca^{2+}. The near-constancy of the extracellular Ca^{2+} concentration is the result of a complex homeostatic system which, in the chicken, primarily involves the secretion of parathyroid hormone (PTH) from the PTG (Brown 1991, Kurokawa 1994). PTH is the major hormone responsible for the minute-to-minute regulation of extracellular Ca^{2+} concentration. The initial event in the expression of PTH bioactivity is binding of the hormone to specific receptors (PTH-Rs) located on the surface of target cells in bone and kidney. The secretion of PTH by the PTG is remarkably sensitive to the extracellular Ca^{2+} concentration, and therefore can be ultraregulated by PTH, which plays a key role in setting the extracellular Ca^{2+} concentration. Changes in the set point of the PTG produce major changes in PTH secretion at any given level of extracellular Ca^{2+} concentration and, in turn, reflect the steady-state level of the plasma Ca^{2+} concentration (Brown 1983). Extracellular Ca^{2+} concentrations therefore modulate cytosolic Ca^{2+} concentrations and other second messengers without actually crossing the cell membrane (Shoback *et al.* 1988, Brown 1991). Cloning of the genes for the avian seven transmembrane domains and G-protein-coupled receptors for Ca^{2+} sensing (CaR; Diaz *et al.* 1997) and for avian PTH (Vortkamp *et al.* 1996) allows, for the first time, more detailed examination of the regulation of the expression of these genes in the chicken.

In this study we summarize our investigations on the regulation of avian CaR gene expression in the PTG, and on PTH-R gene expression in the tibia growth plates of normal chicks and of those afflicted with rickets and tibial dyschondroplasia (TD), as related to changes in plasma Ca^{2+} concentrations.

CaR gene expression in the PTG

The CaR gene was demonstrated to be expressed by the chief cells of the chick PTG by *in situ* hybridization. Immunohistochemistry using a specific chicken PTH antiserum directed against the 1-34 sequence of the chicken hormone revealed that the same chief cells also contain PTH. To elucidate plasma Ca^{2+} regulation of CaR gene expression in the PTG, we evaluated expression of this gene and the PTH content in vitamin D-deficient and Ca-repleted chicks in layers in which massive Ca^{2+} flux occurs, and in chicks in which plasma Ca^{2+} was altered by PTH administration.

Day-old chicks were raised on a vitamin D-deficient diet and transferred to a normal diet at various intervals. As expected, chicks fed a vitamin D-deficient diet had reduced body weight and bone ash content. These reductions were directly related to the length of time on the diet. The concentration of plasma Ca^{2+} peaked after 5 days on the normal diet and leveled off in chicks fed a normal diet for 3 days. The vitamin D-deficient chicks had low levels of CaR mRNA, whereas chicks in the group that exhibited the highest plasma Ca^{2+} concentrations expressed the highest levels of CaRmRNA (Fig. 1). An inverse relationship between CaR gene expression and PTH content was observed: PTH content was higher in glands taken from the vitamin D-deficient chicks than in those from all the repleted groups.

Ca^{2+} turnover in chickens and some other birds during reproduction is more rapid than in any other land vertebrates. During the formation of a calcified egg shell, which contains >10% of the Ca present in the entire chicken's skeleton, plasma Ca^{2+} is completely turned over approximately four times every minute. Maintenance of plasma Ca^{2+} concentration under such extreme conditions is aided by rapid removal of Ca^{2+} from deposits in the medullary bone by the action of PTH. In the PTG of layers during the cycle of eggshell formation, maximal CaR gene expression was found at the time when the egg resides in the eggshell gland where Ca deposition in the shell occurs (Stemberger *et al.* 1977). The lowest levels of PTH in the chief cells were also observed at this time, demonstrating once again that expression of the CaR gene and the PTH content within the PTG chief cells are inversely related and depend on plasma Ca^{2+} concentration.

Administration of a single dose of avian PTH caused a rise in plasma Ca^{2+} concentration which persisted for 1 h at least and then leveled off. No changes in the expression of the CaR gene in the PTG were observed 2 h after PTH administration, whereas after 4 and 6 h, a significant increase in CaR gene expression was observed.

The PTH content in the chief cells was reduced after PTH administration compared with that in the control.

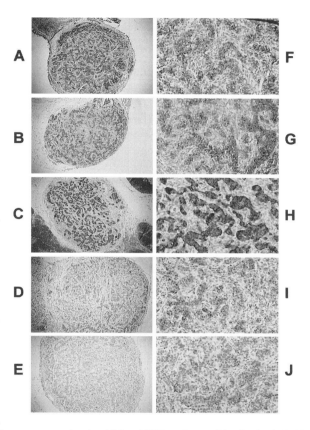

Fig. 1 CaR gene expression in chicken PTG as detected by *in situ* hybridization. The mRNA hybridization signal appears in black in the bright field illumination. Chicks were raised on a vitamin D-deficient diet and were repleted with vitamin D-containing diet as follows: group 1, 10 days of normal diet (A,F); group 2, 7 days of normal diet (B, G); group 3, 5 days of normal diet (C, H); group 4, 3 days of normal diet (D, I). Group 5 was used as a control and received a vitamin D-deficient diet during the entire period (E, J).

PTH-R gene expression in epiphyseal growth-plate chondrocytes in rickets and TD

Longitudinal bone growth occurs as a consequence of proliferation and hypertrophy of growth-plate chondrocytes (Pines & Hurwitz 1991). Chondrocytes located in different regions of the growth plate differ in their differentiation state, morphology, secretion of extracellular matrix components and activities of various enzymes. For example, collagen type II is synthesized by chondrocytes in the proliferative state, whereas alkaline phosphatase (AP) activity and synthesis of collagen type X and osteopontin

(OPN) are restricted to hypertrophic cells (Knopov et al. 1995, 1997, Monsonego et al. 1997, Pines et al. 1999). Growth-plate chondrocytes express the gene coding for the PTH-R (Vortkamp et al. 1996) and respond to PTH with an increase in intracellular cAMP levels (Pines & Hurwitz 1988), a decrease in AP activity (Kato et al. 1990a), and an increase in chondrocyte proliferation (Pines & Hurwitz 1988), leading to inhibition of differentiation (Lee et al. 1996). PTH-related peptide (PTHrP)-depleted mice have shown abnormal epiphyseal cartilage development and increased chondrocyte-programmed cell death (Amizuka et al. 1996). In addition, gene mutation in the PTH-R causes Jansen's chondrodysplasia due to abnormal regulation of the growth-plate chondrocytes and mineral homeostasis (Jüppner 1996), reflecting the important role of PTH in bone growth and metabolism.

To elucidate the involvement of the PTH axis in growth-plate pathology associated with changes in plasma Ca^{2+}, we evaluated PTH-R gene expression in the growth plates of chicks afflicted with rickets. In addition, PTH-R gene expression was evaluated in chicks afflicted with TD in which no changes in plasma Ca^{2+} occur.

In normal chicks, PTH-R gene expression was localized to the lower proliferative and upper hypertrophic zones, the maturation zone of the epiphyseal growth plate. The growth-plate zones were defined by collagen type II and OPN gene expression and AP activity. In addition, PTH-R gene expression occurred in areas surrounding the blood vessels in the hypertrophic zone.

Vitamin D-deficient chicks were characterized by lower growth rate, lower plasma Ca^{2+} and lower bone ash. The proliferating zone of the growth plate of the rachitic chicks was enlarged compared with the matched controls. The growth plate of rachitic animals is characterized by widening and disorganized proliferating and maturation zones and derangement of the cartilaginous cell columns as the result of increased conversion of maturing chondrocytes to hypertrophic chondrocytes (Kato et al. 1990b). In the chicks raised on the vitamin D-deficient diet, the growth plates exhibited the same histology and the same expression of the PTH/PTHrP receptor gene up to day 4 post-hatch. From day 8 onwards, no expression of the PTH/PTHrP receptor gene could be detected (Fig. 2). In rachitic chicks the plasma level of PTH was high compared with the controls. PTH is known to cause downregulation of its receptor functions in a variety of cells such as osteosarcoma (Mitchell & Goltzman 1990), rat osteoblasts (Jorgen et al. 1996), kidney cells (Abou-Samra et al. 1994) and avian epiphyseal growth-plate chondrocytes (Ben-Bassat et al. 1999). Thus the reduction in the PTH/PTHrP gene expression in the rachitic chicks was probably due to this mechanism.

TD was induced by dietary thiram (Ben-Bassat et al. 1999) or by genetic selection for high incidence of lesions (Pines & Hurwitz 1988). No differences in plasma Ca^{2+} were observed between the control and the thiram-treated chicks at all ages examined, although a reduction in the growth rate of the latter was observed. TD is characterized by the appearance of a plug of non-vascularized non-mineralized white opaque cartilage which dominates the proximal metaphysis of the tibiatarsus and occasionally occurs also in the tarsometatarsus (Knopov et al. 1995, 1997, Pines et al. 1999). The

PTH-R gene expression

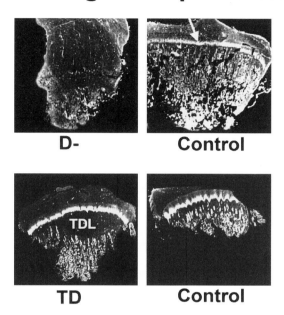

Fig. 2 PTH-R gene expression in the growth plates of chicks afflicted with rickets (D-) or with TD. The mRNA hybridization signal appears in white in the dark field illumination. Chicks were raised on a vitamin D-deficient diet or thiram-containing diet for 17 days. PTH-R gene expression was evaluated by *in situ* hybridization using the avian PTH-R antisense probe. No signal was obtained with the sense probe. Note the disappearance of the PTH-R signal in the vitamin D-deficient chicks and the normal signal in the TD-afflicted chicks compared with the matched controls. TDL, TD lesion.

various morphological and biochemical manifestations of the TD lesion, such as changes in carbonic anhydrase and AP activity, production of collagen types II and X and OPN, suggested that TD chondrocytes failed to undergo the complete differentiation that normally leads to cartilage vascularization and mineralization (Pines *et al.* 1999). In 21-day-old chicks, thiram caused enlargement of the proliferative zone compared with the control growth plates, and the enlargement was dependent on the thiram dose. Normal PTH/PTHrP gene expression was observed at all the ages examined, in the growth plates of the thiram-induced chicks (Fig. 2) and in the growth plates of the chicks selected for high incidence of lesions. These data suggest that alteration in PTH/PTHrP receptor gene expression is involved in the etiology of rickets but not in that of TD.

Concluding remarks

Plasma Ca^{2+} is one of the most efficiently regulated constituents in birds. Its homeostasis is a complex system, involving several components such as various tissues (bone, kidney and intestine), calcitrophic hormones (PTH and 1,25-dihydroxy vitamin D_3) and enzymes (25-cholecalciferol-1-hydroxylase), to name but a few. In addition, feedback mechanisms, the rate of Ca^{2+} fluxes in different compartments, and the time frame of the responses of the various components to alterations in plasma Ca^{2+} concentration also have to be considered. Thus results obtained in experiments performed *in vitro*, *in situ* and *in vivo* should be incorporated into a computer algorithm model (Hurwitz *et al.* 1987) that should be able to predict the events and the outcome in any given physiological state.

References

Abou-Samra AB, Goldsmith PK, Xie LY, Jüppner H, Spiegel AM & Segre GV (1994) Down-regulation of parathyroid (PTH)/PTH-related peptide receptor immunoreactivity and PTH binding in opossum kidney cells by PTH and dexamethasone. *Endocrinology* **135** 2588-2594.

Amizuka, N, Henderson JE, Hoshi K, Warshawsky H, Ozaka H, Goltzman D & Karaplis A (1996) Programmed cell death of chondrocytes and aberrant chondrogenesis in mice homozygous for parathyroid hormone-related peptide gene deletion. *Endocrinology* **137** 5055-5067.

Ben-Bassat S, Genina O, Lavelin I, Leach RM & Pines M (1999) Parathyroid receptor gene expression by epiphyseal growth plates in rickets and tibial dyschondroplasia. *Molecular and Cellular Endocrinology* (In Press).

Brown EM (1983) Four parameter model of the sigmoidal relationship between parathyroid hormone release and extracellular calcium concentration in normal and abnormal parathyroid tissue. *Journal of Clinical Endocrinology and Metabolism* **56** 572-581.

Brown EM (1991) Extracellular Ca^{2+} sensing, regulation of parathyroid cell function, and the roles of Ca^{2+} and other ions as extracellular (first) messengers. *Physiological Reviews* **71** 371-411.

Diaz R, Hurwitz S, Chattopadhyay N, Pines M, Yang Y, Kifor O, Einat MS, Butters R, Herbert SC & Brown EM (1997) Cloning, expression, and tissue.localization of the calcium-sensing receptor in chicken (*Gallus domesticus*). *American Journal of Physiology* **273** R1008-R1016.

Hurwitz S (1989) Calcium homeostasis in birds. *Vitamins and Hormones* **45** 173-221.

Hurwitz S, Fishman S & Talpaz H (1987) Model of plasma calcium regulation: system oscillations induced by growth. *American Journal of Physiology* **252** R1173-R1181.

Jorgen JW, Willemstein-van-Hove EC, Van der Meer JM, Bos MP, Jüppner H, Segre GV, Abou-Samra AB, Feyen JHM & Herrmann-Erlee MPM (1996) Down- regulation of the receptor for parathyroid hormone (PTH) and PTH-related peptide by PTH in primary fetal rat osteoblasts. *Journal of Bone and Mineral Research* **11** 1218-1225.

Jüppner H (1996) Jansen's metaphyseal chondrodysplasia: a disorder due to a PTH/PTHrP receptor gene mutation. *Trends in Endocrinology and Metabolism* **7** 157-162.

Kato Y, Shimazu A, Nakashima K, Suzuki F, Jikko A & Iwamoto M (1990*a*) Effect of parathyroid hormone and calcitonin on alkaline phosphatase activity and matrix calcification in rabbit growth plate chondrocytes cultures. *Endocrinology* **127** 114-118.

Kato Y, Shimazu A, Iwamoto M, Nakashima K, Koike T, Suzuki F, Nishii Y & Sato K (1990*b*) Role of 1,25-dihydroxycholecalciferol in growth-plate cartilage; inhibition of terminal differentiation of chondrocytes *in vitro* and *in vivo*. *Proceedings of the National Academy of Science USA* **87** 6522-6526.

Knopov V, Leach RM, Barak-Shalom T, Hurwitz S & Pines M (1995) Osteopontin gene expression and alkaline phosphatase activity in avian tibial dyschondroplasia. *Bone* **16** 329S-334S.

Knopov V, Hadash D, Hurwitz S, Leach RM & Pines M (1997) Gene expression during cartilage differentiation in turkey tibial dyschondroplasia, evaluated by *in situ* hybridization. *Avian Diseases* **41** 62-72.

Kurokawa K (1994) The kidney and calcium homeostasis. *Kidney International* **45** (Suppl. 44) S97-S105.

Lee K, Lanske B, Karaplis AC, Deeds JD, Kohno H, Nissenson RA, Kronenberg HM & Segre GV (1996) Parathyroid hormone-related peptide delays terminal differentiation of chondrocytes during endochondral bone development. *Endocrinology* **137** 5109-5118.

Mitchell J & Goltzman D (1990) Mechanisms of homologous and heterologous regulation of parathyroid hormone receptors in the rat osteosarcoma cell line UMR-106. *Endocrinology* **126** 2650-2660.

Monsonego E, Baumbach WR, Lavelin I, Gertler A, Hurwitz S & Pines M (1997) Generation of growth hormone binding protein by avian growth plate chondrocytes is dependent on cell differentiation. *Molecular and Cellular Endocrinology* **135** 1-10.

Pines M & Hurwitz S (1988) The effect of parathyroid hormone and atrial natriuretic peptide on cyclic-nucleotide production and proliferation of avian epiphyseal growth plate chondroprogenitor cells. *Endocrinology* **123** 360-365.

Pines M & Hurwitz S (1991) The role of the growth plate in longitudinal bone growth. *Poultry Science* **70** 1806-1814.

Pines M, Knopov V, Genina O, Hurwitz S, Faerman A, Gerstenfeld LC & Leach RM (1999) Development of avian tibial dyschondroplasia: gene expression and protein synthesis. *Calcified Tissue International* **64** (In Press).

Shoback DM, Membreno LA & McGee JG (1988) High calcium and other divalent cations increase inositol triphosphate in bovine parathyroid cells. *Endocrinology* **123** 382-389.

Stemberger BH, Mueller WJ & Leach RM (1977) Microscopic study of the initial stage of eggshell calcification. *Poultry Science* **56** 537-543.

Vortkamp A, Lee K, Lanske B, Segre GV, Kronenberg HM & Tabin C (1996) Regulation of rate of cartilage differentiation by indian hedgehog and PTH-related protein. *Science* **273** 613-622.

Effects of daily administration of human parathyroid hormone (1-34) or salmon calcitonin in green iguanas (*Iguana iguana*)

T J Rosol[1], J L Taylor[1], D G Fischbach[1], V Matkovic[2], M D Eberts[1], S N Huff[1] and K M Morgan[3]

Departments of [1]Veterinary Biosciences and [2]Physical Medicine and Rehabilitation, Ohio State University, 1925 Coffey Road, Columbus, Ohio 43210 and [3]Columbus Zoo, Columbus, Ohio, USA

Introduction

Improper feeding as well as inadequate exposure to UV-B light needed to form vitamin D in captive reptiles leads to a high incidence of secondary hyperparathyroidism (metabolic bone disease) (Boyer 1996), which results in bone loss and pathological fractures. The regulation of calcium (Ca) metabolism is poorly understood in poikilothermic reptiles such as green iguanas (*Iguana iguana*). Treatment of metabolic bone disease is primarily directed at providing available dietary Ca and vitamin D, correcting imbalances in the dietary Ca to phosphorus ratio, and providing a source of UV-B light (Mader 1993, Boyer 1996). The overall goal of this investigation was to study the effects of daily administration of parathyroid hormone (PTH) and calcitonin (CT) on bone mass, bone resorption and formation, and parathyroid gland chief cell morphology in iguanas.

Experimental design and methods

A pilot experiment was conducted in five green iguanas (Glades Herp, Inc., Fort Myers, FL, USA) weighing 165-721 g. They were injected daily for 32 days with salmon CT (Bachem California, Torrance, CA, USA; 7000 IU/mg; 50 IU/kg; 7 µg/kg, $n=2$), human PTH(1-34) (Procter & Gamble; 200 µg/kg, $n=2$) or diluent (saline, pH 5.0, $n=1$). They were fed pelleted Iguana Diet (No 53640600, with vitamin D, 2000 IU/kg, Ca 1.5%, P 0.7%; Zeigier Bros, Gardners, PA, USA) *ad libitum* and were housed (one to two/cage) in metal mesh cages (2.5×4×6 ft) with privacy boxes (two/cage), tree limbs, trays with fresh water and infrared heat lamps (250 W, cage temperature 90-95°F). A Sylvania Gro-Lux lamp (40 W) and UV-B blacklight (40 W) were placed 8 inches above the cages. Body weights ranged from 334 to 1118 g after 32 days. Tail and spine fractures (which did not interfere with ambulation) occurred in the PTH-treated iguanas. The same experimental design was completed with 18 juvenile iguanas (six/group) (96-156 g; The Wild Things, Athens, TN, USA) for 33 days except that the dose of PTH was reduced to 100 µg/kg and calcein (10 mg/kg

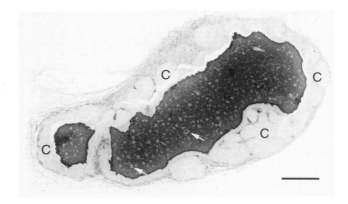

Fig. 1 Parathyroid gland from an iguana administered human PTH(1-34) (200 µg/kg per day) for 1 month. Note the formation of a central follicle-like structure lined by cords of chief cells (C) and containing secretory granule contents and necrotic chief cells (arrows). Bar represents 150 µm.

in saline with 2% $NaHCO_3$) was injected intraperitoneally 1 day before termination. At termination, the iguanas were anesthetized with pentobarbital sodium and blood collected from the heart. Serum Ca was measured by the cresolphthalein complexone method. One parathyroid gland was fixed in 3% glutaraldehyde, post-fixed in 1 M cacodylate buffer and osmium, embedded in medcast plastic, and examined with a Philips 300 electron microscope. The second parathyroid gland, liver, kidneys, heart, and lungs were fixed in 10% neutral-buffered formalin, embedded in paraffin, and examined by light microscopy.

Bone morphometry

Long bones and lumbar vertebrae were fixed in 10% formalin at 4°C for 24-48 h. One proximal tibia was demineralized in 10% EDTA (pH 7.4) for at least 3 days (4°C), infiltrated and embedded in glycol methacrylate (JB-4; Polysciences, Warrington, PA, USA) at 4°C, sectioned at 4 µm, and stained for tartrate-resistant acid phosphatase (Sigma Diagnostics, St Louis, MO, USA; kit No 386-A). Osteoclast number, perimeter and length, osteoblast perimeter, and area of metaphyseal trabecular bone were measured with an Olympus microscope and Bioquant digitized tablet and system IV software (R&M Biometrics, Nashville, TN, USA). Two lumbar vertebrae were embedded in methylmethacrylate, sectioned at 200 µm with a Buehler Isomet low-speed saw, and the percentage of the trabecular perimeter labeled with calcein was measured. One femur from each animal was air-dried and radiographed with a Faxitron X-Ray System (Hewlett-Packard; No 4385SA) to qualitatively evaluate cortical and trabecular bone density. Bone mineral content, density and area were measured with a Lunar densitometer (Lunar, Madison, WI, USA) model DPXL with small animal software (version 1.0e) and a leucite block.

The numerical data are presented as mean ± S.E.M and were analyzed by ANOVA with either the Tukey-Kramer or Student-Newman-Keuls multiple comparison test.

Results

There was no significant change in body weight in the 18 iguanas during the 33 day experiment. Eight iguanas gained up to 13 g and 10 lost up to 16 g. Iguanas injected with CT had a calm demeanor and were easy to catch. In contrast, those injected with PTH were hyperexciteable and difficult to catch. There was little effect of PTH or CT treatment on the histology of the liver, lungs, kidneys or myocardium. One PTH-treated iguana (100 μg/kg per day) had small renal cortical infarcts with tubular mineralization.

There was no significant difference in the total serum Ca concentrations at termination: 2.42±0.22, 2.52±0.40 and 2.50±0.15 mmol/l in control, PTH- and CT-treated iguanas (six/group) respectively.

Histology and ultrastructure of parathyroid gland chief cells

In the pilot experiment there was a marked change in the histology of the parathyroid glands in the PTH-treated (200 μg/kg) iguanas. The center of the glands was replaced by a follicle-like cavity filled with proteinaceous fluid and necrotic chief cells (Fig. 1). Those in the control and CT-treated iguanas were composed of solid cords of chief cells. Ultrastructural examination of the follicles in the PTH-treated iguanas revealed central fluid with the same density as intracytoplasmic secretory granules and contained necrotic chief cells. The chief cells lining the fluid had an accumulation of secretory granules at the cell membrane adjacent to the fluid and secretory granules were undergoing exocytosis into the fluid (Fig. 2). It was interpreted that the majority of the fluid in the center of the glands was composed of secretory granule contents. Chief cells lining the follicle-like structure were joined by multiple intercellular junctions.

Ultrastructure of chief cells

Chief cells from control iguanas contained moderate amounts of rough endoplasmic reticulum and mitochondria with few secretory granules and lysosomes. Chief cells from PTH-treated iguanas had less rough endoplasmic reticulum, increased numbers of secretory and prosecretory granules near the cell membranes and Golgi apparatus, and increased numbers of lysosomes. Chief cells from CT-treated iguanas had prominent interdigitations between cell membranes of adjacent cells, and chief cells were hypertrophied. There was increased rough endoplasmic reticulum, a prominent Golgi apparatus, and moderate numbers of immature and mature secretory granules. Lysosomes were rare. These data were interpreted as showing stimulation of PTH synthesis and secretion by CT treatment and inhibition by PTH treatment.

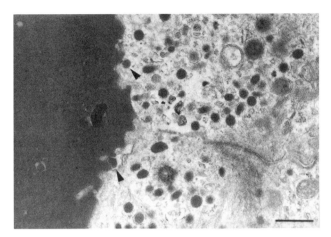

Fig. 2 Electron photomicrograph of chief cells from an iguana administered human PTH(1-34) (200 µg/kg per day) for 1 month. Note the follicle-like structure (left) that contains the contents of secretory granules and the granules that are in the process of being secreted (arrowheads). Bar represents 660 nm.

Bone morphometry

Bone density

No differences in cortical or trabecular bone density were observed on radiographs of the femurs. Bone mineral density (0.102±0.007, 0.096±0.007, 0.100±0.005 g/cm^2) was not different between controls, PTH-treated or CT-treated iguanas.

Osteoclastic bone resorption

There was a 2.7-fold increase ($P<0.01$) in number of osteoclasts/mm of trabecular perimeter and percentage trabecular perimeter lined by osteoclasts in PTH-treated iguanas (100 µg/kg per day) compared with controls (Table 1). There were mild reductions in these parameters in CT-treated iguanas, but the data were not statistically significant. The average length of osteoclasts was similar in the three groups.

Osteoblastic bone formation

Trabecular perimeter lined by active osteoblasts and labeled with calcein was increased 1.7-fold ($P<0.05$) in PTH-treated iguanas compared with controls and CT-treated iguanas (Table 1).

Discussion

Reptiles fed too little Ca and exposed to inadequate levels of UV-B light (285-315 nm) for endogenous vitamin D production develop nutritional secondary hyperparathyroidism (metabolic bone disease) (Allen *et al.* 1996, Boyer 1996). Metabolic bone disease is characterized by increased PTH secretion and stimulation of

Table 1 Effect of daily subcutaneous injection of human PTH (1-34) or salmon CT on trabecular bone in the proximal tibias of green iguanas

Group	Osteoclasts /mm	Osteoclast perimeter (μm)	Osteoclast perimeter (%)	Osteoblast perimeter (%)	Trabecular bone area (%)	Calcein labeled perimeter (%)
Control (n=6)	0.97±0.17	51.5±2.4	5.07±0.96	16.6±2.5	20.3±1.5	17.2±1.8
PTH (n=6)	2.74±0.23**	48.7±2.5	13.20±1.00**	27.4±2.3*	22.2±1.9	30.1±2.9*
CT (n=6)	0.89±0.18	45.0±4.2	4.17±1.01	18.4±2.8	19.4±0.6	13.4±5.4

*$P<0.05$, **$P<0.01$ compared with control.

osteoclastic bone resorption in order to maintain normal serum Ca concentrations (Anderson & Capen 1976). This results in osteopenia and pathological fractures.

Intermittent (once daily) administration of PTH to rats and humans leads to stimulation of bone formation (Canalis *et al.* 1994). The anabolic effects of PTH may be used to treat osteopenia medically. One of the goals of this investigation was to determine whether intermittent administration of PTH would induce bone formation in iguanas. Daily administration of PTH increased both bone resorption and formation in young iguanas. The increased bone formation was probably due to coupled formation in response to increased osteoclastic bone resorption. The anabolic effects of PTH in humans and rats are due to bone formation without prior bone resorption. The data indicate that iguanas respond to PTH differently from mammals and show no anabolic response to intermittent therapy. Iguanas may be very sensitive to the effects of PTH *in vivo* and this may partially explain their high incidence of metabolic bone disease.

It has been recommended that two-weekly doses of CT should be administered to reptiles with metabolic bone disease (Mader 1993). CT inhibits osteoclastic bone resorption, but there are limitations in its clinical use. Osteoclasts become refractory to its effects *in vivo* within hours (Rosol & Capen 1997). After imbalances in diet and UV-B light exposure have been corrected in reptiles with metabolic bone disease, serum PTH levels decline and osteoclastic bone resorption decreases. Therefore, inhibition of osteoclastic bone resorption by CT would be of value temporarily and these animals would benefit from therapy to increase bone mass. Daily administration of CT to normal iguanas for over 1 month was safe, but did not result in a significant increase in bone mass or reduction in osteoclastic bone resorption.

Parathyroid glands from iguanas administered high daily doses of human PTH(1-34) (200 μg/kg per day) developed large central follicles that contained the contents of secretory granules. Follicles with periodic acid/Schiff-positive secretory material have been reported in normal parathyroid glands of lizards (Srivastav et al. 1995). This suggests that iguanas process dispensable PTH differently from other animals. Most mammals degrade unneeded PTH in cytoplasmic lysosomes (Capen 1983). This process probably also occurs in iguanas as lysosomes were present in chief cells. However, they demonstrated a novel mechanism for handling excess PTH. The ultrastructural findings were interpreted as suggesting that unneeded secretory granules and their contents were secreted into a follicle lined by chief cells. The long-term outcome of the follicle contents and whether the contents could be released into the extracellular fluid is not known.

Acknowledgements

This work was supported by the Columbus Zoo and The Ohio State University College of Veterinary Medicine (grant No 734584). Special thanks to Procter & Gamble Pharmaceuticals, Mason, OH, USA (Dr Mark W Lundy) for donating the human PTH and to Dr John Landoll (Department of Physical Medicine and Rehabilitation) for completing the bone densitometry.

References

Allen ME, Oftedal OT & Horst RL (1996) Remarkable differences in the response to dietary vitamin D among species of reptiles and primates: is ultraviolet B light essential? In *Biological Effects of Light,* pp 13-38. Eds MF Hollick & EG Jung. New York: W de Gruyter.

Anderson MP & Capen CC (1976) Nutritional osteodystrophy in captive green iguanas *(Iguana iguana). Virchows Archives Cell Pathology* 21 229-247.

Boyer TH (1996) Metabolic bone disease. In *Reptile Medicine and Surgery,* pp 385-392. Ed DR Mader. Philadelphia: WB Saunders Co.

Canalis E, Hock JM & Raisz LG (1994) Anabolic and catabolic effects of parathyroid hormone on bone and interactions with growth factors. In *The Parathyroids*, pp 65-82. Eds JP Bilezikian, R Marcus & MA Levine. New York: Raven Press.

Capen CC (1983) Structural and biochemical aspects of parathyroid gland function in animals. In *Monograph On Pathology Of Laboratory Animals: Volume 1, Endocrine System, International Life Sciences Institute Series,* pp 217-247. Eds TC Jones, U Mohr & RD Hunt. New York: Springer-Verlag Inc.

Mader DR (1993) Use of calcitonin in green iguanas *(Iguana iguana)* with metabolic bone disease. *Bulletin of the Association of Reptilian and Amphibian Veterinarians* 3 5.

Rosol TJ & Capen CC (1997) Calcium-regulating hormones and diseases of abnormal mineral (calcium, phosphorus, magnesium) metabolism. In *Clinical Biochemistry of Domestic Animals,* pp 619-702. Eds JJ Kaneko, IW Harvey & ML Bruss. San Diego: Academic Press.

Srivastav AK, Sasayama Y & Suzuki N (1995) Morphology and physiological significance of parathyroid glands in reptilia. *Microscopy Research and Technique* 32 91-103.

Influence of dietary calcium and phosphorus on skeletal quality of the modern broiler chicken

B Williams, D Waddington and C Farquharson

Bone Biology Group, Roslin Institute, Roslin, Midlothian EH25 9PS, UK

Introduction

Broiler growth has changed considerably over recent years as a result of genetic selection for meat production; however, poultry diets have changed little in terms of mineral content. Calcium (Ca) and phosphorus (P) are essential for bone formation, and numerous skeletal pathologies are associated with their deficiencies. In a previous study, we compared bone development in a modern fast growing broiler strain with a slower growing genetic precursor (Williams *et al.* 1998). Both strains demonstrated similar periods of rapid bone formation (4-18 days) and mineralisation (4-11 days). However, between the ages of 11 and 18 days the molar Ca:P ratio of cortical bone in the fast growing birds reached 2.15 to 1. This deviation was not shown by the slower growing strain, which consistently displayed molar ratios close to the predicted 1.67 to 1 (Pellegrino & Biltz 1968.)

Nutrient requirements change with age (NRC 1994) and possibly growth of the bird; for example, genetic variation in Ca metabolism has been reported in chickens selected for aspects of body fatness (Shafey *et al.* 1990). The modern strain grew faster and became heavier than its genetic precursor; the size of its bones, and possibly mineral requirements, reflected this. The observed high Ca:P ratio in cortical bone may therefore reflect a dietary inadequacy during this period of rapid bone formation and mineralisation. The present study aimed to investigate this possibility by determining responses in Ca and P metabolism and skeletal health of 2-week-old broilers to differing dietary Ca and available P (avP) contents. It was also hoped that optimum dietary Ca and P contents for maximum skeletal health at this age would be identified.

Methods

Male broilers of the modern strain used previously (Williams *et al.* 1998) were fed diets with a variety of Ca and avP contents in a modified 4×4 factorial design (Fig. 1), with two pens of 15 birds per treatment.

Five birds from each pen were blood-sampled, and culled on day 14. Blood ionised Ca (Ca^{2+}) concentration was measured using a Cibia-Corning 634 Ca^{2+}/pH analyser, and plasma total Ca and inorganic P (P_i) concentrations were determined by colorimetric methods (Rennie *et al.* 1993). Concentrations of markers of bone formation (osteocalcin) and resorption (pyridinoline) were measured by

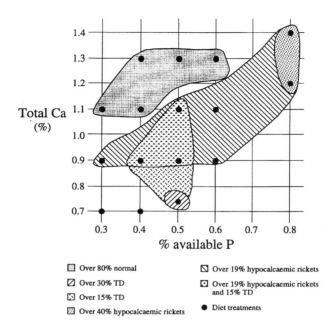

Fig. 1 Diet treatments and predominant pathologies.

radioimmunoassay and HPLC respectively. Blood 1,25-dihydroxyvitamin D_3 (1,25(OH)$_2$D$_3$), was measured using a commercially available kit (IDS, Boldon, Tyne and Wear, UK).

Tibiotarsus breaking strength was measured using a Lloyd Instruments LRX material tester. Proximal tibiotarsus growth plates were assessed for rickets and tibial dyschondroplasia (TD) by histology. A segment of bone was taken from the same point on each shaft for ashing and measurement of mineral (%ash), Ca (%Ca) and P (%P) contents by atomic absorption spectrophotometry (Varian AA-875 series) and colorimetry (Technicon Traacs 800) respectively.

For all variables, models based on a quadratic response surface in dietary Ca, avP and their product were used to identify an optimum diet. Linear responses with Ca and avP were also examined. The proportion of birds with particular growth plate pathologies was analysed as binomial variates on the logistic scale.

Results

Optimum diet

Growth plate assessment indicated that diets should contain 1.1-1.3%Ca and 0.3-0.6%avP for normal development (Fig. 1). Although the observed maxima of some bone quality and growth measures did coincide, the optimum diet varied with the variable investigated. Similarly the response surfaces also failed to identify a single optimum diet, and for many individual variables no clear optimum was identified.

Pathology

The incidence of TD was markedly increased by reducing Ca ($P<0.001$) and by increasing avP ($P<0.01$). The incidence of hypocalcaemic rickets increased with avP content along a band of dietary Ca ($P<0.05$). The proportion of birds with normal growth plates therefore increased with dietary Ca ($P<0.001$) and at lower avP contents ($P<0.001$; see Fig. 1). Hypophosphataemic rickets occurred too rarely for analysis to be performed.

Blood and plasma measures

All subsequent results are linear responses. Concentrations of plasma $1,25(OH)_2D_3$, pyridinoline and total Ca were unaffected by diet. Birds given higher dietary avP showed greater concentrations of plasma P_i ($P<0.05$) and osteocalcin ($P<0.01$), and lower concentrations of Ca^{2+} ($P<0.001$; Fig. 2). Concentrations of plasma osteocalcin were also lower at higher dietary Ca content ($P<0.05$).

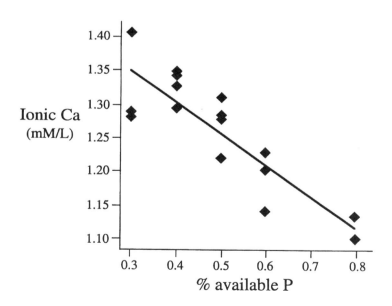

Fig. 2 Effect of dietary available phosphorus on plasma ionised Ca.

Bone quality

There were no clear relationships of bone breaking strength, %ash, %Ca or %P with dietary Ca or avP content. However, the molar Ca:P ratio in cortical bone showed a marked decline as dietary avP increased ($P<0.05$). All diets gave high Ca:P ratios, and individual values ranged from 1.82 to 3.89.

Discussion

Normal plasma Ca concentration is controlled by the action of three hormones: parathyroid hormone (PTH), $1,25(OH)_2D_3$ and calcitonin (Taylor & Dacke 1984). In the present experiment, $1,25(OH)_2D_3$ and total plasma Ca concentrations were unaffected by dietary Ca or avP content, whilst Ca^{2+} concentration was lower in birds fed on diets with higher avP. Excess dietary Ca or P may reduce the availability of the other by precipitation in the intestine (Shafey et al. 1990); however, the unaltered plasma total Ca concentrations suggest that such problems were not encountered here. The lower Ca^{2+} concentrations seen in birds fed high avP diets do not necessarily contradict this. Ca circulates in three fractions: ionised, protein-bound and that complexed to anions, including P_i (Bushinsky & Monk 1998). Plasma P_i concentration was greater in birds given the diets with higher avP content, and it is conceivable that at least some of this extra P_i was complexed with Ca in the plasma, thereby lowering Ca^{2+}.

Production of $1,25(OH)_2D_3$ would be expected to increase under conditions producing low Ca^{2+} concentration by the action of increased circulating PTH on renal $25(OH)D_3$ 1-hydroxylase; however, this was not seen here. At present we have no estimates of 'normal' Ca^{2+} for the broiler strain used, and it is possible that low dietary avP actually results in high Ca^{2+} concentrations. Calcitonin may become an important Ca regulator under such conditions, and it would therefore be of interest to measure its circulating concentration.

TD incidence increased on the lower Ca and higher avP diets, supporting other observations (Rennie et al. 1993) that incidence of TD is influenced by dietary Ca:P ratio. The incidence of hypocalcaemic rickets also increased at higher avP contents along a band of increasing dietary Ca, upholding the results of Thorp (1994) who suggested that the dietary balance of Ca and P was important. The increased incidence of both disorders in birds given higher dietary avP is associated with the lower circulating Ca^{2+}, strongly suggesting that hypocalcaemia was occurring. Overall, pathological assessment of the growth plate indicated that diets should be high in Ca (1.1-1.3%) and low in avP (0.3-0.6%) to give normal cartilage development.

Bone resorption was unaffected by dietary Ca and avP content in this study. However, plasma osteocalcin concentration was greater in birds on diets higher in avP and lower in Ca, a feature that agrees with the results of Corlett & Care (1988). The higher osteocalcin concentration may be a consequence of the low plasma Ca^{2+} in birds fed these diets. $1,25(OH)_2D_3$ stimulates osteocalcin production by osteoblasts and, whilst osteocalcin is a marker for bone formation, it has recently been implicated in the recruitment and differentiation of bone-resorbing cells (Lian & Marks 1990).

However, as $1,25(OH)_2D_3$ concentrations were unaffected by diet, these relationships require further investigation.

One of the main concerns about bone quality is the likelihood of fractures in birds of market age (6 weeks). Our results show that tibiotarsus breaking strength was unaffected by dietary Ca and avP content at 2 weeks of age. However, it is possible that a dietary effect may only become apparent later in the birds' development.

Despite complex effects on bone mineral content, a relatively simple dietary effect was seen on the molar Ca:P ratio – higher dietary avP giving smaller cortical bone Ca:P ratios. This was probably caused by the lower circulating Ca^{2+} and greater P_i concentration produced by the higher avP diets. All diets gave bone Ca:P ratios higher than the expected 1.67 to 1, values ranging from 1.82-3.89 to 1. This was surprising because avP contents in some diets were almost double those currently recommended. In addition, although increasing avP in the diet gave ratios closer to the expected 1.67 to 1, it also resulted in poor growth plate development and would not therefore be desirable. The reason for this result is at present unclear, but a Ca:P ratio of up to 2.97 to 1 does not seem to preclude relatively good quality bone.

Acknowledgements

Many thanks to Ross Breeders Ltd for providing birds, and to Ross Breeders Ltd, John Thomson and Sons Ltd, and MAFF for funding. Thanks also to J Williams, Y Nys and colleagues (INRA, France) for help with osteocalcin analysis, to A Bailey and colleagues at Bristol University for pyridinoline analysis, and to Drs C Whitehead and M McLeod at the Roslin Institute for nutritional advice.

References

Bushinsky DA & Monk RD (1998) Calcium. *Lancet* **352** 306-311.

Corlett SC & Care AD (1988) The effects of reduced dietary phosphate intake on plasma osteocalcin levels in sheep. *Quarterly Journal of Experimental Physiology* **73** 443-445.

Huyghebaert G (1996) Effects of dietary calcium, phosphorus, Ca/P-ratio and phytase on zootechnical performances and mineralisation in broiler chicks. *Archiv fur Geflügelkunde* **61** 53-61.

National Research Council (1994) *Nutrient Requirements of Poultry*, edn 9. Washington DC: National Academy Press.

Lian JB & Marks Jr SC (1990) Osteopetrosis in the rat: coexistence of reductions in osteocalcin and bone resorption. *Endocrinology* **126** 995-962.

Pellegrino ED & Biltz RM (1968) Bone carbonate and the Ca to P molar ratio. *Nature* **219** 1261-1262.

Rennie JS, Whitehead CC & Thorp BH (1993) The effect of dietary 1,25-dihydroxychole calciferol in preventing tibial dyschondroplasia in broilers fed on diets imbalanced in calcium and phosphorus. *British Journal of Nutrition* **69** 809-816.

Shafey TM, McDonald MW & Pym RAE (1990) The effect of dietary calcium upon growth rate, food utilisation and plasma constituents in lines of chickens selected for aspects of growth or body composition. *British Poultry Science* **31** 577-586.

Taylor TG & Dacke CG (1984) Calcium metabolism and its regulation. *Physiology and Biochemistry of the Domestic Fowl*, vol 5, pp125-170. Ed BM Freeman. London: Academic Press.

Thorp B (1994) Skeletal disorders in the fowl: a review. *Avian Pathology* **23** 203-236.

Williams B, Solomon S, Waddington D, Thorp B & Farquharson C (1998) Determining the life history of broiler bone mineralisation. *British Poultry Science* **39** (Suppl.) 59-60.

Measurement of nitric oxide production from isolated single chick osteoclasts using a porphyrinic microsensor

S F Silverton[1], O A Adebanjo[2], B S Moonga[2], G D Markham[3], T Malinski[4], J V Johnson[5] and M Zaidi[2]

[1]School of Dental Medicine, University of Pennsylvania, [2]Department of Medicine, Medical College of Pennsylvania, Hahnemann School of Medicine, and Center for Osteoporosis and Skeletal Aging, Geriatrics and Extended Care Service, Philadelphia Veterans Affairs Medical Center; [3]Fox Chase Cancer Institute, Philadelphia, PA; [4]Department of Chemistry, Oakland University, Rochester, MI; [5]Department of Chemistry, University of Florida, Miami, FL, USA

NO and bone resorption: an overview

NO is a powerful *in vivo* and *in vitro* inhibitor of osteoclastic bone resorption (MacIntyre *et al.* 1991, Oursler *et al.* 1991, Zaidi *et al.* 1993, Sunyer *et al.* 1997). The osteoclast is also a major producer of NO (Silverton *et al.* 1995). Both forms of NO synthase (NOS), eNOS and iNOS, are present in the osteoclast (Brandi *et al.* 1995, Sunyer *et al.* 1997). The catalytic activity of eNOS is induced by both Ca^{2+} and protein kinase C (Sunyer *et al.* 1997). Osteoclastic NO production could therefore be triggered by changes in cytosolic $[Ca^{2+}]$, such as in response to an elevated extracellular $[Ca^{2+}]$, hence the hypothesis that inhibitory effects of Ca^{2+} on bone resorption are exerted through NO release, possibly via iNOS induction. Thus, as expected, inhibitors of NOS, namely aminoguanidine and N^G-monomethyl-L-arginine, attenuate the effects of extracellular Ca^{2+} on bone resorption (Brandi *et al.* 1995, Sunyer *et al.* 1997).

However, to obtain insights into NOS regulation in the osteoclast, the effect of Ca^{2+} on iNOS expression must be separated from its effect on eNOS catalysis. This is not possible with current NO detection methods which depend on slow accumulation of NO by-products, such as nitrite and citrulline, to threshold levels (Green *et al.* 1982, Bredt & Snyder 1989). To distinguish these, we have made real-time NO measurements using a microsensor (Malinski & Taha 1992) and compared this with two other real-time methods, namely quadrupole ion trap mass spectrometry (QITMS) and electron paramagnetic resonance (EPR) spectroscopy with spin trapping (Zweier *et al.* 1995). Of these, the microsensor method proved most reliable and sensitive and was used in subsequent studies. Furthermore, because of the inherent heterogeneity of our primary cultures and because the released NO is oxidized rapidly during diffusion, we have sampled NO as close as possible to the osteoclast (Zaidi *et al.* 1988).

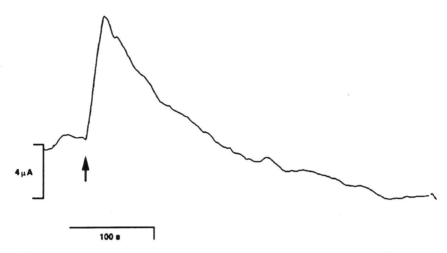

Fig. 1 Representative microsensor current traces (μA) in response to application (at ↑) of the NOS cofactor NADPH (5 μM) to chick osteoclasts.

Porphyrinic microsensor measurements of NO

The NO microsensor

NO production from single osteoclasts was measured electrochemically using a porphyrinic microsensor. The latter was prepared by depositing a film of polymeric porphyrinic (Ni(II) tetrakis-(3-methoxy-4-hydroxyphenyl)porphyrin) on a thermally sharpened carbon fiber electrode coated with Nafion (Malinski *et al.* 1996). NO detection by the microsensor depends upon the fast and selective oxidation of NO to NO^+ followed by rapid NO^+ removal by the Nafion ion-exchange layer. The sensor current density ranged between 1.5 and 1.8 mA/cm^2. It was stabilized by differential pulse voltammography so that resulting current changes were linearly proportional to NO concentration changes at the microsensor surface. The detection limit of the NO microsensor approached 10^{-9} M.

Microsensor calibration

Before each experiment, we prepared a standard saturated solution of NO (2 mM) in an O_2-depleted buffer (200 mM NaH_2PO_4/Na_2HPO_4, pH 7.4) under helium. The microsensor was placed in 2 ml Hanks balanced salt solution (HBSS; from Gibco) or Moscona's low-Ca solution in the well of a microscope (Diaphot; Nikon, Tokyo, Japan). The reference electrode was placed in the same well and the rectifying signal stabilized. Several 2 μl aliquots of the standard NO solution were drawn up in a Hamilton gas-tight syringe and added sequentially to the well. The peak height for each current signal observed was averaged within every experiment and for every electrode used.

There was a linear relationship between the peak microsensor current and added NO concentration. We could thus calculate the NO concentration of a given sample by plotting its peak current value compared with the NO standard. In each experiment, we used one or more electrodes and calibrated each against a standard. This allowed us to make comparisons of NO concentrations across experiments. Interestingly, the current amplitude of the saturated NO standard, in nA/µM NO (median 13.7; range 2.4-17.8; $n=5$), was markedly lower than that obtained with NO in O_2-depleted buffer (median 137; range 25-254; $n=6$). We next examined the effect of increasing the distance between the point of NO application and the microsensor. When this was increased from 2 to 4 mm, the sharp monophasic peaks were converted to significantly attenuated biphasic signals that followed up to a ~60 s lag (not shown). Taken together, these results reflect NO degradation by O_2 and other buffer constituents in our open-well system.

NO measurement from single osteoclasts

After a 16 h incubation in humidified CO_2 (5%) at 37°C, chick osteoclasts plated on dentine or glass were placed in either HBSS or Moscona's low-calcium buffer

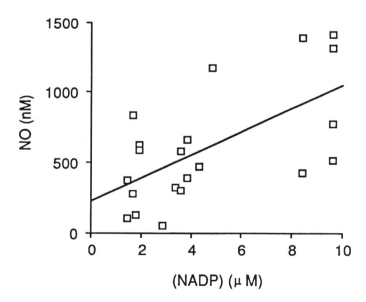

Fig. 2 Correlation between calculated NO concentration (nM) and applied NADPH concentration (µM) ($r=0.625$, $P=0.0024$, $n=21$ measurements).

(pH 7.4). A micromanipulator was used to position the microsensor in close apposition to the dorsal surface of one or more osteoclasts. NO measurements were made after addition of NADPH (1-10 µM), L-arginine (1 mM) and superoxide dismutase (1 KU/l). Between four and eight osteoclast-dentine or osteoclast-glass cultures were sampled for each preparation.

NO transients in isolated osteoclasts were elicited through the application of NADPH (1-10 µM) (Silverton *et al.* 1989). Each transient consisted of a rapid rise to a peak value followed by an exponential decay to near-basal levels (Fig. 1). NADPH serves as a cofactor for NOS, and at submaximal catalytic rates, the peak height is expected to be a function of the NADPH concentration. Thus, we found a significant concentration-dependent correlation ($r=0.625$; $n=21$, $P=0.0024$) between released [NO] and applied [NADPH] (Fig. 2). Finally, to confirm that the NO signal resulted from NADPH action on NOS, we used N-ω-nitro-L-arginine methyl ester (L-NAME), a non-selective NOS inhibitor. We found that 300 µM-L-NAME inhibited the NO signal significantly (Fig. 3).

Oxidation of NO by the microsensor occurs several orders of magnitude more rapidly than for most other organic and inorganic species (Malinski *et al.* 1996). Such species, therefore, generally do not interfere with NO detection. An exception is aminoguanidine, another non-selective NOS inhibitor. The latter, at 250 µM, triggered a large current surge that was irreversible. In addition, L-arginine (1 mM), superoxide

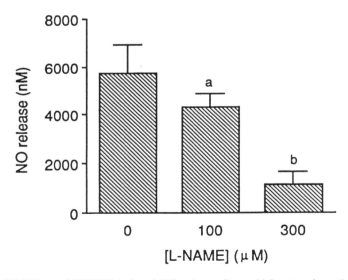

Fig. 3 Inhibition of NADPH-induced NO release from chick osteoclasts (mean± S.E.M.) by the non-selective NOS inhibitor, L-NAME (100 and 300 µM). [a]$P<0.05$, [b]$P<0.01$ compared with control (no L-NAME).

dismutase (1 kU/l) and thapsigargin (2 µM) all do not affect the microsensor current (not shown).

Calculations of cellular NO release were made from the peak current tracing. Its rate of rise was used to derive the rate of [NO] change, in nM/min; the latter was then converted to a release rate, in nmol/min. For this, we first defined high and low estimates for NO detection using volumes around the osteoclast and microsensor, respectively. We then used these estimates to calculate NO release rates (Table 1). The latter were up to ~10-fold higher than those derived from other cells (Ralston *et al.* 1994).

Table 1 NO release from osteoclasts using two estimates of volume around the osteoclast and around the microsensor electrode.

	Around cell	Around electrode
Volume (m^3)	6×10^{-13}	2×10^{-12}
NO release (pmol/min per 10^7 cells)	1728	518

A major strength of the microsensor is in its ability to measure free radicals at or near their production site. This is particularly relevant to labile osteoclast-derived free radicals, such as NO and O_2^-, which are destroyed rapidly in the intracellular, extracellular and plasma membrane compartments. For example, while purified neutrophil-derived myeloperoxidase produces ~30 amol O_2^-/s per cell^{-1} (0.1 unit myeloperoxidase per 2.5×10^7 cells), neutrophils themselves only produce 0.126 amol O_2^-/s per cell (Steibeck *et al.* 1994). Although, corresponding data have not yet been obtained for NO, we know that NO signal intensity is attenuated markedly as the application distance from the microsensor is increased. Thus, by applying the microsensor closest to the osteoclast surface, we are, in effect, enhancing specificity and taking advantage of the negligible diffusional distance.

Acknowledgements

These studies were supported by grants from the National Institutes of Health (R29-AR20248 to SFS and RO1-AG141917 to MZ) and, in part, by Department of Veterans Affairs (Merit Review Award to MZ). The authors are grateful to Professor Iain MacIntyre, MD, FRS, for extensive discussions.

References

Brandi ML, Hukkanen M, Umeda T, Moradi-Bidhendi N, Bianchi S, Gross SS, Polak JM & MacIntyre I (1995) Bidirectional regulation of osteoclast function by nitric oxide synthase isoforms. *Proceeding of the National Academy of Science of the USA* **92** 2954-2958.

Bredt S & Snyder SH (1989) Nitric oxide mediates glutamate-linked enhancement of cGMP levels in the cerebellum. *Proceeding of the National Academy of Science of the USA* **86** 9030-9033.

Green L C, Wagner D A, Glogowski J, Skipper P L, Wishnok JS & Tannenbaum SR (1982) Analysis of nitrate, nitrite, and [15N] nitrate in biological fluids. *Analytical Biochemistry* **126** 131-138

MacIntyre I, Zaidi M, Alam ASMT, Datta HK, Moonga BS, Lidbury PS, Hecker M & Vane J (1991) Osteoclastic inhibition: an action of nitiric oxide not mediated by cyclic GMP. *Proceeding of the National Academy of Science of the USA* **88** 2936-2940

Malinski T, Mesaros S & Tomboulian P (1996) Nitric oxide measurement using electrochemical methods. *Methods in Enzymology* **268** 58-69

Malinski T & Taha Z (1992) Nitric oxide release from a single cell measured *in situ* by a porphyrinic-based microsensor. *Nature* **358** 676-678

Oursler MJ, Collin-Osdoby P, Anderson F, Li L, Webber D & Osdoby P (1991) Isolation of avian osteoclasts: improved techniques to preferentially purify viable cells. *Journal of Bone & Mineral Research* **6** 375-385

Ralston SH, Todd D, Helfrich M, Benjamin N & Grabowski PS (1994) Human osteoblast-like cells produce nitric oxide and express inducible nitric oxide synthase. *Endocrinology* **135** 330-336

Silverton SF, Matsumoto H, DeBolt K, Reginato A & Shapiro I M (1989) Pentose phosphate shunt metabolism by cells of the chick growth cartilage. *Bone* **10** 45-51

Silverton SF, Mesaros S, Markham GD & Malinski T (1995) Osteoclast radical interactions: NADPH causes pulsatile release of NO and stimulates superoxide production. *Endocrinology* **136** 5244-5247

Steinbeck MJ, Appel WHJr, Verhoeven AJ & Karnovsky MJ (1994) NADPH-oxide expression and *in situ* production of superoxide by osteoclasts actively resorbing bone. *Journal of Cell Biology* **126** 765-772

Sunyer T, Rothe L, Kirsch D, Jiang X, Anderson F, Osdoby P & Collin-Osdoby P (1997) Ca^{2+} or phorbol ester but not inflammatory stimuli elevate inducible nitric oxide synthase messenger ribonucleic acid and nitric oxide (NO) release in avian osteoclasts: autocrine NO mediates Ca^{2+} inhibited bone resorption. *Endocrinology* **136** 2148-2162

Zaidi M, Alam ASMT, Bax BE, Shankar VS, Bax CMR, Gill JS & Pazianas M (1993) Role of the endothelial cell in osteoclast control: new perspectives. *Bone* **14** 97-102

Zaidi M, Chambers TJ, Bevis PJR, Beacham JL, Gaines Das RE & MacIntyre I (1988) Effects of peptides from the calcitonin genes on bone and bone cells. *Quarterly Journal of Experimental Physiology* **73** 471-485

Zweier JL, Wang P & Kuppusamy P (1995) Direct measurement of nitric oxide generation in the ischemic heart using electron paramagnetic resonance spectroscopy. *Journal of Biological Chemistry* **270** 304-307

Variation in the concentration and binding affinities of the plasma vitamin D-binding protein

C J Laing[1,2], G M Shea[2] and D R Fraser[1]

[1]Department of Animal Science and [2]Department of Veterinary Anatomy and Pathology, University of Sydney, Australia 2006

Introduction

Vitamin D and its metabolites are known to associate with a number of non-specific carriers in the bloodstream, including lipoproteins (Rikkers & DeLuca 1967, Hay & Watson 1976), chylomicrons (Dueland et al. 1982) and albumin (Bikle et al. 1985). However, early work showed that vitamin D associated with the lipoprotein fraction of blood plasma will, over time, shift to the α-globulin pool until the majority of the vitamin D is associated with this protein fraction (Rikkers & DeLuca 1967). A protein has since been identified that acts as a highly specific binding protein for vitamin D compounds (Cooke & Haddad 1989).

It is now thought that vitamin D-binding protein (DBP) represents the most divergent member of a multigene family which includes albumin and α-fetoprotein (McLeod and Cooke 1989). It has been estimated that DBP diverged from the common albumin/α-fetoprotein ancestor some 600 million years ago (Haefliger et al. 1989). Not surprisingly a specific protein carrier for vitamin D compounds has been identified in all major vertebrate phyla except cartilaginous fish (Hay & Watson 1976, Van Baelen et al. 1988, Licht 1994). It has been estimated that the DBP gene is the slowest evolving of the DBP/ALB/AFP group (Haefliger et al. 1989), and a high rate of nucleotide conservation has been reported among mammals (Braun et al. 1993, Osawa et al. 1994).

DBP in the serum of mammals and chickens has been demonstrated to possess a single high-affinity binding site for its ligands (Haddad & Walgate 1976, Bouillon et al. 1978, 1980). Although only a single DBP has been identified to bind all vitamin D compounds in mammals (Edelstein et al. 1973, Haddad & Walgate 1976, Bouillon et al. 1978), Hay & Watson (1976), in a survey of 22 fish, 12 amphibia, five reptile and 19 bird species, found two binding proteins in the plasma of the chicken, turkey, goose and dove. It has since been suggested that, in the chicken, this second DBP, an oestrogen-dependent post-translational modification of the plasma DBP, is involved specifically with the transport of vitamin D into the egg (Edelstein et al. 1973, Fraser and Emtage 1976). This is probably also the case with the other species of birds in which Hay and Watson have identified multiple DBP bands, as these were the only species capable of year-round egg production. A second vitamin D-binding fraction has also been identified in the alligator (Licht 1994), despite not being found in the previous study by Hay and Watson, which also supports the view that vertebrates possess a single DBP, which may be modified under certain circumstances.

Equilibrium exists between DBP-bound and unbound ligand in the bloodstream, and it is the unbound ligand that is available for cellular uptake (Bikle *et al.* 1985). The two properties known to influence the proportion of ligand that is specifically bound and that which is unbound are the concentration of the DBP in the bloodstream and the affinity with which it binds its ligand. These properties regulate the rate of metabolic or catabolic conversion and the rate of utilisation of the vitamin D compounds in the body (Vargas *et al.* 1990, Bouillon *et al.* 1996).

Rikkers &DeLuca (1967) were the first to suggest that DBP may possess different affinities for distinct forms of vitamin D. It has since been shown in mammals that DBP possesses greatest affinity for 25-hydroxyvitamin D_3 ($25(OH)D_3$), followed by 1,25-dihydroxyvitamin D_3 ($1,25(OH)_2D_3$) and then vitamin D_3 (Cooke & Haddad 1989). However, it is also known that the absolute affinity of DBP for various ligands is species dependent (Vieth *et al.* 1990).

We were interested in characterising the specific binding properties of plasma (and hence of DBP) for $25(OH)D_3$, and investigating whether variation in the binding characteristics, and hence in the calculated proportion of free $25(OH)D_3$, occurs among reptile and bird species.

Methods

Blood plasma samples from a wide variety of mainly indigenous Australian birds and reptiles were obtained, and came from both wild-caught animals and animals housed in captivity in various zoological parks.

Plasma DBP concentration and its dissociation constant for $25(OH)D_3$ were calculated from maximum binding of increasing concentrations of tritiated 25-hydroxyvitamin D_3 ($[^3H]25(OH)D_3$) under equilibrium conditions, assuming a single class of specific binding proteins, with a single ligand-binding site per molecule. The method is a modification of one described elsewhere (Woloszczuk 1985). The data were analysed by a non-linear regression which describes the combined specific and non-specific binding of ligand, and derives the specific component mathematically (Swillens 1995). This technique was chosen as a large percentage of the $[^3H]25(OH)D_3$ is bound by the DBP, which makes the direct measurement of non-specific binding unreliable. The 'free 25(OH)D index' was calculated as the ratio of K_d to DBP concentration, and is an indication of the proportion of 25(OH)D not bound to DBP.

Because of variation in distribution and variance between the groups, non-parametric tests were used to investigate differences. The Kruskall-Wallis test was used to assess distribution differences between groups, and Mann-Whitney tests were used to determine significant differences between pairs. Significance was inferred from a value of $P<0.05$, except when multiple Mann-Whitney tests were performed, when significance was inferred at $P<0.01$.

Results and discussion

Values for DBP concentration, K_d and 'free 25(OH)D index' are listed in Table 1. Kruskall-Wallis tests confirmed significant differences among the species studied, for each of the three variables.

We were interested in investigating whether this variation was consistent with phylogenetic history, along the principle that greater variation may be expected between lineages that diverged earlier in vertebrate history, than between more recently diverged groups. In order to investigate this, the species were assigned to one of three major phylogenetic lineages (see Table 1) – Lepidosauria (snakes, lizards and tuataras), Chelonia (turtles and tortoises) and Archosauria (crocodilians and birds) – and the lineages compared, using mean values for species as data points.

There were significant differences in DBP concentration among species within the archosaurs ($P<0.001$), and within the lepidosaurs ($P<0.001$). There was no significant difference in the DBP concentration between the two chelonians ($P=0.27$). There were no significant differences between lineages in DBP concentration ($P=0.23$), nor specifically between the DBP values for archosaurs and lepidosaurs ($P=0.13$).

Significant differences were found in K_d for 25(OH)D among archosaurs ($P=0.001$), lepidosaurs ($P<0.001$) and between the chelonians ($P=0.0015$). Although a significant difference was observed in average K_d values between the three lineages, the significance level was an order of magnitude less powerful than that demonstrated within each lineage ($P=0.015$). Further, there was a significant difference between the distribution of mean K_d values of lepidosaurs and archosaurs, although also at a lower significance level than that shown within each lineage ($P=0.019$).

Significant differences were detected in 'free index' for 25(OH)D among archosaurs ($P<0.001$) and lepidosaurs ($P<0.001$), although there was no significant difference in this variable between the two chelonians. No significant difference was observed in the distribution of mean 'free index' values between the three lineages ($P=0.17$), nor specifically between archosaurs and lepidosaurs ($P=0.27$).

For all three variables a greater variation was detected within each lineage than between the lineages, which suggests that variation in DBP is not occurring at a high phylogenetic level, but rather at lower levels, within families and genera. This was confirmed by pairwise analysis of closely related species within the lineage Lepidosauria. The species within this study that comprise Lepidosauria belong to the following groups: snakes (*M. spilota, S. suta, N. scutatus* and *B. irregularis*), varanid lizards (*V. varius* and *V. gouldii*), scincid lizards (*E. stokesii, E. whitii, T. rugosa, Eulamprus* spp. and *C. robustus*), gekkonid lizards (*H. binoei, G. variegata* and *O. tryoni*), and agamid lizards (*P. vitticeps, C. nuchalis* and *P. lesueurii*). Significant differences ($P<0.01$) were found in DBP concentration between at least one pair each of snakes and skinks, in K_d between at least one pair each of snakes, skinks and geckos, and in the 'free index' of 25(OH)D between at least one pair each of skinks and geckos. However, when these groups were compared using mean values for species as data points, no significant differences were found for DBP concentration,

Table 1 Plasma analyses on reptile and bird species from three major phylogenetic lineages. Values are mean ± S.D.

Species	N	[DBP] (μmol/L)	Kd (nM)	Free 25(OH)D
Archosauria				
King quail (*Coturnix chinensis*)	10	1.83 ± 0.52	0.17 ± 0.09	0.10 ± 0.07
Pekin duck (*Anas platyrhynchos*)	6	3.00 ± 0.91	0.15 ± 0.11	0.05 ± 0.02
Black cockatoo (*Calyptorhynchus banksii*)	5	2.21 ± 1.15	0.09 ± 0.03	0.05 ± 0.01
Lorikeet (*Trichoglossus haematodus*)	5	0.49 ± 0.25	0.98 ± 1.08	2.39 ± 2.30
Gull (*Larus* sp.)	3	1.67 ± 1.13	0.29 ± 0.12	0.21 ± 0.16
Little penguin (*Eudyptula minor*)	7	1.74 ± 0.52	0.21 ± 0.06	0.14 ± 0.09
Magpie (*Gymnorhina tibicen*)	3	0.55 ± 0.30	0.16 ± 0.12	0.28 ± 0.12
Java finch (*Padda oryzivora*)	4	0.48 ± 0.05	0.06 ± 0.01	0.12 ± 0.04
Alligator (*Alligator mississippiensis*)	4	0.97 ± 0.30	0.73 ± 0.14	0.77 ± 0.08
Chelonia				
Aldabran tortoise (*Geochelone gigantea*)	8	1.81 ± 0.99	0.97 ± 0.37	0.60 ± 0.20
Broad shelled turtle (*Chelodina expansa*)	6	2.82 ± 1.15	1.92 ± 0.53	0.88 ± 0.82
Lepidosauria				
Tuatara (*Sphenodon* sp.)	9	0.90 ± 0.31	0.35 ± 0.29	0.37 ± 0.22
Bearded dragon (*Pogona vitticeps*)	9	2.86 ± 1.02	0.59 ± 0.32	0.22 ± 0.11
Netted dragon (*Ctenophorus nuchalis*)	7	1.41 ± 0.66	0.40 ± 0.18	0.36 ± 0.26
Water dragon (*Physignathus lesueurii*)	3	2.69 ± 1.66	0.66 ± 0.24	0.27 ± 0.06
Tree dtella (*Gehyra variegata*)	7	1.90 ± 0.35	0.55 ± 0.18	0.30 ± 0.13
Bynoe's gecko (*Heteronotia binoei*)	6	1.42 ± 0.66	1.07 ± 0.26	1.02 ± 0.72
Spotted velvet gecko (*Oedura tryoni*)	4	2.04 ± 0.35	1.33 ± 0.51	0.65 ± 0.21
Gidgee (*Egernia stokesii*)	4	2.47 ± 0.55	0.73 ± 0.13	0.32 ± 0.15
White's skink (*Egerna whitii*)	5	4.06 ± 1.25	0.41 ± 0.12	0.10 ± 0.02
Shingleback lizard (*Tiliqua rugosa*)	6	1.38 ± 0.54	0.28 ± 0.07	0.22 ± 0.09
Striped skink (*Ctentous robustus*)	7	4.06 ± 0.78	0.34 ± 0.13	0.09 ± 0.04
Water skink (*Eulamprus* spp.)	6	1.94 ± 1.16	1.13 ± 0.18	0.78 ± 0.44
Lace monitor (*Varanus varius*)	7	1.14 ± 0.41	1.34 ± 0.58	1.25 ± 0.45
Gould's monitor (*Varanus gouldii*)	10	1.68 ± 0.80	1.05 ± 0.27	0.80 ± 0.53
Diamond python (*Morelia spilota*)	7	2.88 ± 0.57	0.38 ± 0.32	0.13 ± 0.10
Curl snake (*Suta suta*)	7	1.08 ± 0.40	0.07 ± 0.04	0.07 ± 0.04
Tiger snake (*Notechis scutatus*)	7	1.89 ± 0.82	0.20 ± 0.07	0.13 ± 0.08
Brown tree snake (*Boiga irregularis*)	3	1.70 ± 0.72	0.31 ± 0.10	0.19 ± 0.04

and although significant differences were determined for K_d ($P=0.04$) and 'free index' ($P=0.04$), the significance was not as powerful as that determined within each group. This supports the finding that the largest variations in vitamin D-binding characteristics occur at the species level, rather than between broader phylogenetic groupings.

Although it is not possible from these data to explain what factors are responsible for determining these differences between closely related species, the most likely explanation is that they represent an adaptive response to ecological conditions.

Conclusions

These data demonstrate that variation does occur in the characteristics of the plasma DBP, and in the resultant functional component, the 'free index', among reptile and bird species. Furthermore, they demonstrate that the variation does not reflect broad phylogenetic groupings. It would appear that the differences observed in these binding characteristics do not represent random variation based on evolutionary distance, but are more likely the result of adaptation at species level to ecological circumstances. This supports our view of DBP as a protein with an active, rather than a simple transporting, role in the physiological economy of vitamin D.

Acknowledgements

The authors wish to thank Angelika Trube for technical assistance. We would also like to acknowledge the staff at Taronga Zoo, Sydney, the Royal Melbourne Zoo, the Australian Museum, Sydney, and Dr Alison Cree, Department of Zoology, University of Otago, for providing plasma samples.

References

Bikle DS, Siiteri P, Ryzen E & Haddad J (1985) Serum protein binding of 1,25-dihydroxy vitamin D: a reevaluation by direct measurement of free metabolite levels. *Journal of Clinical Endocrinology and Metabolism* **61** 969-975.

Bouillon R, Van Baelen H, Rombauts W & DeMoor P (1978) The isolation and characterisation of the vitamin D-binding protein from rat serum. *Journal of Biological Chemistry* **253** 4426-4431.

Bouillon R, Van Baelen H, Tan B & DeMoor P (1980) The isolation and characterisation of the 25(OH)-vitamin D-binding protein from chick serum. *Journal of Biological Chemistry* **255** 10925-10930.

Bouillon R, Verstuyf A, Zhao J, Tan B & Van Baelen H (1996) Nonhypercalcemic vitamin D analogs: interactions with the vitamin D-binding protein. *Hormone Research* **45** 117-121.

Braun A, Kofler A, Morawietz S & Cleve H (1993) Sequence and organization of the human vitamin D-binding protein gene. *Biochimica et Biophysica Acta* **1216** 385-394.

Cooke N &. Haddad J (1989) Vitamin D binding protein (Gc globulin). *Endocrine Reviews* **10** 294-304.

Dueland S, Pederson JI, Helgerund P & Drevon CA (1982) Transport of vitamin D_3 from rat intestine. *Journal of Biological Chemistry* **257** 146-150.

Edelstein S, Lawson D & Kodicek E (1973) The transporting proteins of cholecalciferol and 25-hydroxycholecalciferol in serum of chicks and other species. *Biochemical Journal* **135** 417-426.

Fraser DR & Emtage JS (1976) Vitamin D in the avian egg. *Biochemical Journal* **160** 671-682.

Haddad J & Walgate J (1976) 25-Hydroxyvitamin D transport in human plasma. *Journal of Biological Chemistry* **251** 4803-4809.

Haefliger D, Moskaitis J, Schenberg D & Wakli W (1989) Amphibian albumins as members of the albumin, alpha-foetoprotein, vitamin D-binding protein multigene family. *Journal of Molecular Evolution* **29** 344-354.

Hay A & Watson G (1976) The plasma transport proteins of 25-hydroxycholecalciferol in fish, amphibians, reptiles, and birds. *Comparative Biochemistry and Physiology* **53B** 167-172.

Licht P (1994) The relation of the dual thyroxine/vitamin D-binding protein (TBP/DBP) of Emydid turtles to the vitamin D-binding proteins of other vertebrates. *General and Comparative Endocrinology* **94** 215-224.

McLeod J F & Cooke NE (1989) The vitamin D-binding protein, α-fetoprotein, albumin multigene family: detection of transcripts in multiple tissues. *Journal of Biological Chemistry* **264** 21760-21769.

Osawa M, Tsuji T, Yukawa N, Saito T & Takeichi S (1994) Cloning and sequence analysis of cDNA encoding rabbit vitamin D-binding protein (Gc-globulin). *Biochemistry and Molecular Biology International* **34** 1003-1009.

Rikkers H & DeLuca H (1967) An in vivo study of the carrier proteins of ^{3}H-vitamins D_3 and D_4 in rat serum. *American Journal of Physiology* **213** 380-386.

Swillens S (1995) Interpretations of binding curves obtained with high receptor concentrations: practical aid for computer analysis. *Molecular Pharmacology* **47** 1197-1203.

Van Baelen H, Allewaert K & Bouillon R (1988) New aspects of the plasma carrier protein for 25-hydroxycholecalciferol in vertebrates. *Annals of the New York Academy of Sciences* **538** 60-68.

Vargas S, Bouillon R, Van Baelen H & Raisz L (1990) Effects of vitamin D-binding protein on bone resorption stimulated by 1,25-dihydroxyvitamin D_3. *Calcified Tissue International* **47** 164-168.

Vieth R, Kessler M & Pritzher K (1990) Species differences in the binding kinetics of 25-hydroxyvitamin D_3 to vitamin D-binding protein. *Canadian Journal of Physiology and Pharmacology* **68** 1368-1371.

Woloszczuk W (1985) Determination of vitamin D binding protein by Scatchard analysis and estimation of a free 25-hydroxy-vitamin D index. *Clinica Chimica Acta* **145** 27-35.

Japanese quail medullary bone, an *in vivo* model for bone turnover: effect of disodium pamidronate

C G Dacke, J Sanz, K Foster, J Anderson and J Cook

School of Pharmacy and Biomedical Sciences, University of Portsmouth, Portsmouth PO1 2DT, UK

Introduction

Medullary bone, a highly labile calcium reservoir, is found within the endosteal cavities of long bones in egg-laying hens. It forms shortly before the onset of egg laying and persists throughout the egg-laying period, its formation being concomitant on the maturation of the ovarian follicles. It is the most overtly oestrogen-sensitive form of vertebrate bone and as such is totally dependent on the presence of gonadal steroids for its formation and shows rapid phases of formation and resorption in seasonal birds (Dacke 1979, 1999). Furthermore, it can be induced within a few days in adult male birds such as Japanese quail, by injecting them with oestrogenic steroids (Miller & Bowman 1981, Schraer & Hunter 1985). After withdrawal of the oestrogens, the medullary bone resorbs with equal rapidity (Dacke *et al.* 1993). This makes it a potentially useful model for studies of drugs such as the bisphosphonates, used to treat human postmenopausal osteoporosis. Preliminary studies indicate that the bisphosphonate alendronate can protect structural bone and inhibit medullary bone formation if given to chickens before the commencement of egglay. When given during egglay the drug reduces medullary bone volume and, at higher doses, eggshell quality (Thorp *et al.* 1993). The aim of the present study was to investigate the ability of a bisphosphonate, disodium pamidronate, to influence both formation and the subsequent resorption of medullary bone.

Methods

Two experiments were carried out. In both, medullary bone formation was induced in 8- to 10-week-old adult male quail (110-130g) by injecting them subcutaneously with three doses of 400 µg 17β-oestradiol dissolved in 200 µl corn oil at 3 day intervals. In the first experiment the birds received three doses of 0.03 (low dose) or 0.12 mg (high dose) disodium pamidronate in 200 µl distilled water at 2 day intervals for 6 days before the start of oestrogen dosage; injections of water used as carrier served as control. In the second experiment, the birds were treated with the bisphosphonate for 6 days immediately after dosage with oestrogen. They were anaesthetised with ether, exsanguinated by cardiac puncture and killed by cervical dislocation 3 days after the final injection of oestrogen or bisphosphonate. Blood samples were assayed for total and ionised Ca. Long bones (femur and tibia) were removed, decalcified, processed

Table 1 Effect of disodium pamidronate on medullary bone formation and resorption.

	Expt 1 (Pre-oestrogen treatment with bisphosphonate)			Expt 2 (Post-oestrogen treatment with bisphosphonate)		
	Vehicle	Low-dose bisphosphonate	High-dose bisphosphonate	Vehicle	Low-dose bisphosphonate	High-dose bisphosphonate
Medullary bone area (%)	12.28±1.0	8.18±2.01	5.34±1.39*	2.70±0.86	5.12±1.94*	6.31±2.23***
Cortical bone area (mm^2)	1.22±0.04	1.29±0.08	1.25±0.06	1.12±0.08	1.21±0.14	1.61±0.07
Femur ash weight (mg)	169±20	154±10	149±10*	137±10	159±5	168±8*
Plasma total Ca (mM)	9.06±1.25	9.41±2.41	13.14±1.51*	6.38±1.10	5.00±0.67	5.81±0.26
Plasma Ca^{2+} (mM)	1.46±0.02	1.47±0.02	1.53±0.02*	1.03±0.04	0.96±0.04	1.01±0.09

All data are mean ± s.E.M. Significance versus appropriate control group was determined by one-way ANOVA followed by Tukey-Kramer post-hoc test; *$P<0.05$, ***$P<0.001$ ($n=7-12$ animals per group).

into paraffin wax, sectioned into 6 μm slices and stained with haematoxylin and eosin. Other bones were ashed at 600°C overnight. In the sectioned bones total medullary bone area (excluding cellular components) was quantified microscopically by using an eyepiece graticule. For this purpose a graticule with 100 crosses in the visual field was used. A positive score was given where a cross coincided with an area of medullary bone. For each bird, three sections from each femur (taken from one third the distance from the top of the shaft) were quantified by this method and the mean values for these three sections used. This allowed an accurate estimation of the area within the endosteal cavity containing medullary bone. We were unable to quantify mineralised area in undecalcified sections because of the extremely brittle nature of medullary bone. Cortical bone area was quantified by conventional image analysis.

Results

Pretreatment with bisphosphonate resulted in a significant and dose-dependent reduction in the percentage area of the endosteal cavity containing medullary bone (Table1). Treatment with the bisphosphonate after oestrogen treatment resulted in a significant reduction in medullary bone resorption. These values were reflected to a limited extent by significant variation in bone ash weights for the various treatment groups. Cortical bone area was not significantly affected by any treatment. There were no significant differences in either total or ionised plasma Ca after the various treatments. Total plasma Ca levels were higher and plasma ionised Ca levels were lower in experiment 1 than experiment 2.

Discussion and conclusions

The results indicate that treatment of male Japanese quail with a bisphosphonate can inhibit the formation of medullary bone which occurs after treatment with oestrogen for a few days. If the bisphosphonate is given to animals in which medullary bone has been induced by oestrogen treatment, then it inhibits the ensuing resorption of medullary bone. These responses are extremely rapid compared with those in mammals and can be detected within a few days of treatment with the bisphosphonate. Cortical bone does not appear to be markedly affected by the bisphosphonate. Plasma ionised and total Ca levels were higher in bisphosphonate-treated animals than controls and appear to be inversely correlated with medullary bone area in experiment 1, while the effect was much less marked and not significant in experiment 2. Both total and ionised Ca levels were higher in experiment 1 than experiment 2, probably reflecting the more imminent exposure to oestrogen in experiment 1. In birds, high oestrogen levels result in high plasma Ca levels because of Ca binding by yolk protein precursors (Dacke 1979). In these experiments it was not possible to differentiate between mineralised medullary bone and non-mineralised osteoid, as attempts to cut undecalcified sections of medullary bone were unsuccessful because of the highly brittle nature of this bone. However, the fact that medullary bone areas were reflected to some extent by bone ash weights indicates that there are differences in medullary bone mineral content as well as apparent total bone area. It is concluded that treatment with a bisphosphonate can inhibit both medullary bone formation and resorption. The responses are rapid and apparent within 15 days of the start of treatment.

References

Dacke CG (1979) *Calcium Regulation in Sub-Mammalian Vertebrates*. London: Academic Press.

Dacke CG (1999) The parathyroids, calcitonin and vitamin D. In *Sturkie's Avian Physiology*, 5th edn. Ed G Causey Whittow. Orlando: Academic Press (In Press).

Dacke C, Arkle S, Cook D, Wormstone I, Jones S, Zaidi M & Bascal Z (1993) Medullary bone and avian calcium regulation. *Journal of Experimental Biology* **184** 63-84.

Miller SC & Bowman BM (1981) Medullary bone osteogenesis following estrogen administration to mature male Japanese quail. *Developmental Biology* **87** 52-63.

Schraer H & Hunter S (1985) The development of medullary bone: a model for osteogenesis. *Comparative Biochemistry and Physiology* **82** 13-17.

Thorp B H, Wilson S, Rennie S & Solomons S (1993) The effect of a bisphosphonate on bone volume and eggshell structure in the hen. *Avian Pathology* **22** 671-682.

Concentrations of 1,25-dihydroxyvitamin D_3 and its plasma-binding protein are related to genetic susceptibility to tibial dyschondroplasia in the chicken

M Lowe[1], C J Laing[1,2], W L Bryden[1] and D R Fraser[1]

[1]Department of Animal Science, and [2]Department of Veterinary Anatomy and Pathology, University of Sydney, NSW, Australia 2006

Introduction

Tibial dyschondroplasia (TD) is a disease of interrupted chondrocytic differentiation and bone mineralisation at the epiphysis and is characterised grossly by an avascular cartilaginous plug in the proximal tibial growth plate. Leach & Nesheim (1965) first described this cartilage abnormality in the broiler chicken; however, the underlying mechanism is still to be fully elucidated.

Leach & Nesheim (1965) demonstrated the significance of the genetic predisposition to TD when they were able to develop broiler strains with either a high incidence (41%) or low incidence (16%) of TD after just one generation of selection. Further studies confirmed the heritability of TD with selective breeding (Leach & Nesheim 1972, Riddell 1976). Furthermore, there are some breeds of chicken in which TD is rarely or never seen even when fed TD-inducing diets (Edwards 1984).

The expression of TD is influenced by nutrition, and vitamin D has been implicated in its aetiology. Dietary 25-hydroxyvitamin D_3 ($25OHD_3$) and 1,25-dihydroxyvitamin D_3 ($1,25(OH)_2D_3$) have been shown to decrease the incidence of the disease in birds fed a TD-inducing diet (Edwards 1989). Susceptibility to TD appears to be related to differences in metabolism of vitamin D_3. Vitamin D_3 and its metabolites are transported in the bloodstream in association with a specific high-affinity carrier, the vitamin D-binding protein (DBP; Cooke & Haddad 1989). The characteristics of the plasma transport of vitamin D_3 and its metabolites are important in mediating the rate of metabolism and utilisation of these compounds, as it is only the fraction not associated with DBP that is available for cellular uptake (Bikle et al. 1985).

In this study the concentrations of vitamin D metabolites and their DBPs were investigated in two commercial strains of broiler chickens with differing susceptibilities to TD. The aim was to determine whether the mechanism of $1,25(OH)_2D_3$ transport is involved in the pathogenesis of the disease.

Methods

Two broiler strains, demonstrated to differ in their susceptibility to TD in this laboratory, were obtained from two commercial hatcheries. Strain A was the susceptible strain and strain B was tolerant to TD induction.

Male chicks of both strains were raised in pens and fed a wheat/sorghum-based diet which was formulated to be nutritionally adequate (National Research Council 1994). The diet had been shown previously not to influence the incidence of TD and was fed *ad libitum* throughout the experiment (27 days). At the end of the experiment, birds were bled via cardiac puncture and tibias were macroscopically examined for signs of dyschondroplasia. All procedures were performed according to ACEC regulations.

Plasma concentrations of the vitamin D metabolites, $25OHD_3$ and $1,25(OH)_2D_3$, were measured using standard competitive protein-binding assays (Mason & Posen 1977, Reinhardt & Hollis 1986). DBP concentration was measured indirectly using a saturation binding system which utilised increasing concentrations of $[^3H]25OHD_3$. Specific and non-specific binding were determined from experimental binding curves using a non-linear regression model (Swillens 1995). Plasma total Ca concentrations were determined using atomic absorption spectrophotometry.

Data distribution for each variable was investigated using the Anderson-Darling test, and homogeneity of data was tested with Bartlett's (normal data) or Levene's (non-normal data) tests. Differences between groups was assessed using Student's *t*-tests (normal distribution) or Mann-Whitney *U*-test (non-parametric analysis).

Results

No dyschondroplastic lesions were observed in the tibial growth plates of any of the birds. Plasma DBP concentration, $1,25(OH)_2D_3$ concentration, and the index of free $1,25(OH)_2D_3$ (as estimated by the ratio of ligand to DBP) were found to be significantly higher ($P<0.05$) in strain A birds than birds of strain B at 27 days of age (Table 1). Plasma total calcium concentration was also found to be significantly higher ($P<0.05$) in strain A broilers. There were no differences in the plasma concentration of $25OHD_3$ between the two strains of birds, nor were the final body weights of the birds significantly different.

Discussion

Several authors have reported differences in susceptibility to the disease among commercially available broiler strains. While Veltmann & Jensen (1981) were unable to show any differences in susceptibility to TD among nine broiler strain crosses, Elliot & Edwards (1994) found that 'Peterson×Hubbard chicks' had a significantly higher incidence and severity of TD than two other broiler strains tested and had a higher $1,25(OH)_2D_3$ plasma concentration in three of four experiments.

In the present experiment, higher blood concentrations of $1,25(OH)_2D_3$ were also measured in the more susceptible strain. Furthermore, as none of the birds in the current study developed clinical TD, it is concluded that the increased $1,25(OH)_2D_3$

Table 1 A comparison of plasma vitamin D and calcium concentrations in two strains of broiler chickens with a high (A) and low (B) susceptibility to TD

Variable	Strain A	Strain B
DBP (µmol/l)	2.33 ± 0.34^a (26)	2.05 ± 0.28^b (18)
1,25(OH)$_2$D$_3$ (pmol/l)	298 ± 122^a (8)	168 ± 60^b (8)
Ca (mmol/l)	1.22 ± 0.23^a (24)	1.05 ± 0.12^b (17)
25OHD$_3$ (nmol/l)	43 ± 13 (23)	37 ± 15 (17)

Values are mean±S.D. (n).
Different superscript letters within a row denote significant differences ($P<0.05$).

concentrations in the susceptible strains are not a response to the disease process but could be involved in the pathogenesis.

That disease incidence and severity can be overcome by increasing dietary intake of vitamin D suggests that a derangement in the metabolism and/or utilisation of vitamin D compounds may be involved in the pathogenesis of TD. Elliot & Edwards (1994) suggested that differences in susceptibility to TD among broiler chickens may lie in their ability to utilise 1,25(OH)$_2$D$_3$ at the receptor level. There is evidence that the activity of 1,25(OH)$_2$D$_3$ receptors is upregulated by 1,25(OH)$_2$D$_3$ (Costa et al. 1985). This upregulation occurs at the protein and mRNA level (Pike 1991). It has been proposed that, in birds susceptible to TD, the receptor may not be as sensitive to autoinduction by plasma 1,25(OH)$_2$D$_3$.

However, as higher plasma total calcium concentrations were observed in the more susceptible birds, the calcaemic activity of the 1,25(OH)$_2$D$_3$ does not appear to be affected. This implies that the expression/function of the vitamin D receptor in both strains was adequate to increase calcium concentrations. There was no evidence in this study for a decreased sensitivity of the vitamin D receptor in strain A. A simple difference in the vitamin D receptor gene may not fully explain the differing genetic susceptibility to TD. However, the possibility that differential expression in chondrocytes may be involved cannot be discounted.

The higher DBP concentration in strain A indicates that this protein may play a role in the pathogenesis of TD. The similarity in 25OHD$_3$ concentrations between the two strains suggests that the role of DBP in the disease process is more likely to involve the mediation of the metabolism and/or utilisation of 1,25(OH)$_2$D$_3$.

Acknowledgements

The authors would like to gratefully acknowledge the support of the Chicken Meat Committee of the Rural Industries Research and Development Corporation. We would like to thank Angelika Trube for her technical assistance.

References

Bikle DS, Siiteri P, Ryzen E & Haddad J (1985) Serum protein binding of 1,25-dihydroxyvitamin D: a reevaluation by direct measurement of free metabolite levels. *Journal of Clinical Endocrinology and Metabolism* **61** 969-975.

Cooke N & Haddad J (1989) Vitamin D binding protein (Gc globulin). *Endocrine Reviews* **10** 294-304.

Costa EM, Hirst M & Feldman D (1985) Regulation of 1,25-dihydroxyvitamin D_3 receptors by vitamin D analogs in cultured mammalian cells. *Endocrinology* **117** 2203-2210.

Edwards HM Jr (1984) Studies on the etiology of tibial dyschondroplasia. *Journal of Nutrition* **114** 1001-1003.

Edwards HM Jr (1989) The effect of dietary cholecalciferol, 25 hydroxycholecalciferol and 1,25-dihydroxycholecalciferol on the development of tibial dyschondroplasia in broiler chickens in the absence and presence of disulfiram. *Journal of Nutrition* **119** 647-652.

Elliot MA & Edwards HM Jr (1994) Effect of genetic strain, calcium, and feed withdrawal on growth, tibial dyschondroplasia, plasma 1,25-dihydroxycholecalciferol, and plasma 25-hydroxycholecalciferol in sixteen-day-old chickens. *Poultry Science* **73** 509-519.

Leach RM Jr & Nesheim MC (1965) Nutritional, genetic and morphological studies of an abnormal cartilage formation in young chicks. *Journal of Nutrition* **86** 236-244.

Leach RM Jr & Nesheim (1972) Further studies on tibial dyschondroplasia (cartilage abnormality) in young chicks. *Journal of Nutrition* **119** 647-652.

Mason RS & Posen S (1977) Some problems associated with assay of 25-hydroxycholecalciferol in human serum. *Clinical Chemistry* **23** 806-810.

National Research Council (1994) *Nutrient Requirements of Poultry*. Washington, DC: National Academy of Sciences.

Pike JW (1991) Vitamin D_3 receptors: structure and function in transcription. *Annual Review of Nutrition* **11** 189-216.

Reinhardt TA & Hollis RL (1986) 1,25-Dihydroxyvitamin D microassay employing radioreceptor techniques. *Methods in Enzymology* **123** 176-185.

Riddell C (1976) Selection of broiler chicken for a high and low incidence of tibial dyschondroplasia with observations on spondylolisthesis and twisted legs (perosis). *Poultry Science* **55** 145-151.

Swillens S (1995) Interpretations of binding curves obtained with high receptor concentrations: practical aid for computer analysis. *Molecular Pharmacology* **47** 1197-1203.

Veltmann JR & Jensen LS (1981) Tibial dyschondroplasia in broilers: comparison to dietary additives and strains. *Poultry Science* **60** 1473-1478.

Parathyroid hormone and estrogen effects on adhesion of chicken medullary bone osteoclasts

T Sugiyama[1,2], S Kusuhara[1] and C V Gay[2]

[1]Department of Animal Science, Niigata University, Niigata 950-21, Japan and
[2]Department of Biochemistry and Molecular Biology, The Pennsylvania State University, University Park, Pennsylvania 16803, USA

Introduction

Medullary bone is specific to female birds and plays an important role as a calcium reservoir for eggshell calcification. On the medullary bone surface, osteoclasts undergo cyclic functional modifications during the egg-laying cycle. More specifically, when an egg is in the magnum of the oviduct, osteoclasts cease resorbing bone and profound bone formation occurs; therefore this phase is referred to as the bone-formative phase. When an egg is in the shell gland of the oviduct, osteoclasts resorb medullary bone and supply calcium for eggshell calcification; this phase is termed the bone-resorptive phase (Sugiyama & Kusuhara 1993). Osteoclastic bone resorption associated with the egg-laying cycle is regulated by systemic calcium-regulating hormones such as parathyroid hormone (PTH), estrogen (E2) and calcitonin (Gay 1988, Dacke *et al.* 1993).

Bone resorption is a multistep process, involving the adhesion of osteoclasts to mineralized bone surfaces followed by development of ruffled borders. The $\alpha v \beta 3$ integrin is expressed on the osteoclast cytoplasmic membrane and provides a way for osteoclasts to recognize bone matrix-residing arginine-glycine-asparatic acid (RGD) motifs contained in osteopontin, collagen and bone sialoprotein (Horton *et al.* 1991, Greenfield *et al.* 1991, Flores *et al.* 1992). Osteoclast adhesion to osteopontin via $\alpha v \beta 3$ integrin results in cytoskeletal modifications and induction of bone resorption (Teitelbaum *et al.* 1995).

In the present study, we isolated homogenous inactive and active osteoclasts from chicken medullary bone at the bone-formative and bone-resorptive phases, and estimated the ability of the isolated osteoclasts to adhere to bone slices. Additionally, the effects of PTH and E2 on osteoclast adhesion capacity were examined; involvement of RGD-sequence recognition by PTH-and E2-treated osteoclasts was also evaluated.

Materials and methods

Isolation and purification of osteoclasts for adhesion studies

The procedure for the culture of medullary bone osteoclasts was slightly modified from the original method (Zambonin-Zallone *et al.* 1982). In brief, medullary bone

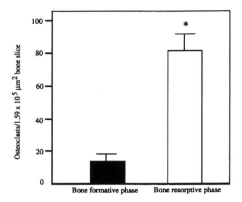

Fig. 1 Number of osteoclasts isolated from two medullary bone phases adhering to bone slices. Results are mean±S.D. *$P<0.01$ compared with bone-formative phase.

was dissected from the femurs of egg-laying chickens at the bone-formative and bone-resorptive phases, and then were gently scraped and crushed in PBS to obtain a cell suspension rich in osteoclasts. Osteoclasts were isolated from the cell suspension by unit gravity sedimentation in PBS+10% BSA. Finally, the isolated osteoclasts were resuspended in phenol red-free minimal essential medium (α-MEM) supplemented with 10% heat-inactivated fetal bovine serum (FBS) and antibiotics, and cultured on bovine bone slices for 24 h. After culturing, tartrate-resistant acid phosphatase (TRAP)-positive cells were counted as osteoclasts.

Effects of PTH and E2 on the adhesion capacity of osteoclasts

A cell suspension rich in osteoclasts was obtained from chicken medullary bone as described above, except that purification by unit gravity sedimentation was omitted. The osteoclast-rich cell suspension was cultured on bone slices in 10% FBS-containing MEM with human PTH (0.4U/ml; Asahikasei, Tokyo, Japan), 17β-estradiol (E2; 1×10^{-8} M; Sigma, St Louis, MO, USA) or vehicle for 24 h. After culturing, adherent TRAP-positive osteoclasts were counted. The adhesion capacity of osteoclasts was represented as the percentage of attached osteoclasts that had been treated compared with vehicle-treated cells. Cells were also cultured in FBS-free MEM with GRGDSP peptide (($0-25.6)\times10^{-3}$ M; Gibco, Grand Island, NY, USA), an inhibitor of RGD-dependent adhesion, and the numbers of osteoclasts adhering to the bone slices was determined.

Results

As shown in Fig. 1, osteoclasts from the bone-formative phase adhered to bone slices at an average of 12.12 cells/1.59×10^5 µm² whereas those isolated during the bone-resorptive phase adhered to the bone slices more frequently (82.30 cells/1.59×10^5 µm²).

The effects of PTH and E2 on adhesion are shown in Fig. 2. The adhesion capacity of osteoclasts from the bone-formative phase was not significantly altered (90±4%) when treated with PTH but was reduced to 51±16% of control levels when treated with E2. The adhesion capacity of osteoclasts from the bone-resorptive phase was stimulated to 149±10% of control levels by PTH, whereas a non-statistically significant change was detected (79±19%) for E2-treated cells. Fig. 3A shows that PTH-stimulated adhesion capacity of osteoclasts was inhibited to control levels by RGD peptides (−59%). In contrast, the decreased adhesion of E2-treated osteoclasts was only slightly influenced by RGD peptides (−16%). The adhesion capacity of vehicle-treated osteoclasts was substantially inhibited by RGD-containing peptides, being reduced to that of E2-inactivated osteoclasts (Fig. 3B). The non-RGD-containing control peptide, GRGESP, did not affect the adhesion of osteoclasts (data not shown).

Discussion

Previous studies have demonstrated that there are distinct differences in the distribution and function of $\alpha v \beta 3$ integrin between active and inactive osteoclasts (Lakkakorpi *et al.* 1989, Holt & Marshall 1998). In the present study, the adhesion capacity of osteoclasts isolated from medullary bone in the bone-resorptive phase was high, while that of isolated osteoclasts from the bone-formative phase was low. The results indicate that the adhesion capacity of osteoclasts varies in relation to egg-laying cycle phases and correlates with cell function.

Additionally, E2 inhibited the adhesion capacity of the osteoclasts from the bone-formative phase, whereas PTH stimulated adhesion of cells from the bone-resorptive phase. These results suggest that the capacity of medullary bone

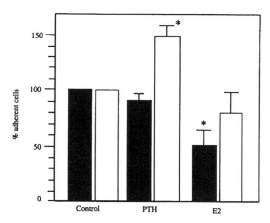

Fig. 2 Effects of calcium-regulating hormones, PTH (0.4 U/ml) and E2 (1×10^{-8}M) on the adhesion of medullary bone osteoclasts isolated from the bone-formative (solid bars) and bone-resorptive (open bars) phases. Results are mean±S.D. *$P<0.01$ compared with control.

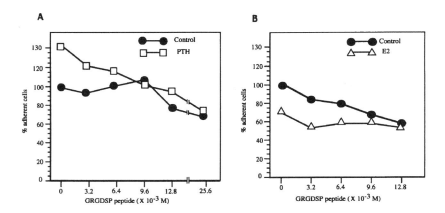

Fig. 3 Effects of RGD competition on adhesion of osteoclasts in the presence of (A) 0.4 U/ml PTH (osteoclasts from bone-resorptive phase) and (B) 1×10^{-8}M E2 (osteoclasts from the bone-formative phase).

osteoclasts to adhere to bone is regulated by PTH and E2. Also, the present studies with RGD-peptides indicate that osteoclastic recognition of the RGD-sequence motifs, a component of bone matrix, can be influenced by these hormones. However, the adhesion of osteoclasts was not entirely inhibited by RGD-peptides, suggesting that the primary adhesion of osteoclasts could depend on additional adhesion factors (Lakkakorpi *et al.* 1991, Ilvesaro *et al.* 1998).

Acknowledgements

Supported by a Research Fellowship from the Japan Society for the Promotion of Science for Young Scientists and NIH Grant No. DE 09459.

References

Dacke CG, Arkle S, Cook DJ, Wormstone IM, Jones S, Zaidi M & Bascal ZA (1993) Medullary bone and avian calcium regulation. *Journal of Experimental Biology* **184** 63-88.

Flores ME, Norgard M, Heinegard D, Reinholt FP & Anderson G (1992) RGD-directed attachment of isolated rat osteoclasts to osteopontin, bone sialoprotein, fibronectin. *Experimental Cell Research* **201** 526-530.

Gay CV (1988) Avian bone resorption at the cellular level. *CRC Critical Reviews in Poultry Science* **1** 197-210.

Greenfield EM, Teti A, Grano M, Colucci S, Zambonin Zallone A, Ross FP, Teitelbaum SL, Cheresh D & Hruska KH (1991) Recognition of osteopontin and related peptides by integrin $\alpha v \beta 3$ stimulates immediate cell signals in osteoclasts. *Journal of Biological Chemistry* **30** 20369-20374.

Holt I & Marshall MJ (1998) Integrin subunitβ3 plays a crucial role in the movement of osteoclasts from the periosteum to the bone surface. *Journal of Cellular Physiology* **175** 1-9.

Horton MA, Taylor ML, Arnett TR & Helfrich MH (1991) Arg-Gly-Asp (RGD) peptides and the anti-vitronectin receptor antibody 23C6 inhibit dentine resorption and cell spreading by osteoclasts. *Experimental Cell Research* **195** 368-375.

Ilvesaro JM, Lakkakorpi PT & Vaananen HK (1998) Inhibition of bone resorption *in vitro* by a peptide containing the cadherin cell adhesion recognition sequence HAV is due to prevention of sealing zone formation. *Experimental Cell Research* **242** 75-83.

Lakkakorpi P, Tuukanen J, Flenkunen T, Yarvelin K & Vanaanen HK (1989) Organization of osteoclast microfilaments during the attachment to bone surface *in vivo*. *Journal of Bone and Mineral Research* **4** 817-825.

Lakkakorpi PT, Morton MA, Helfrich MH, Karhukorpi E-K & Vaananen HK (1991) Vitronectin receptor has a role in bone resorption but does not mediate tight sealing zone attachment of osteoclasts to the bone surface. *Journal of Cell Biology* **115** 1179-1186.

Sugiyama T & Kusuhara S (1993) Ultrastructural changes of osteoclasts on hen medullary bone during the egg-laying cycle. *British Poultry Science* **34** 471-477.

Teitelbaum SL, Abu-Amer Y & Ross FP (1995) Molecular mechanisms of bone resorption. *Journal of Cellular Biochemistry* **59** 1-10.

Zambonin-Zallone A, Teti A & Primavera MV (1982) Isolated osteoclasts in primary culture: first observations on structure and survival in culture media. *Anatomy and Embryology* **165** 405-413.

Parathyroid gland hyperplasia in female snapping turtles (*Chelydra serpentina*) during egg-laying

T J Rosol[1], P C Stromberg[1], J L Taylor[1], S W Fisher[2] and J F Estenik[3]

[1]Departments of Veterinary Biosciences and [2]Entomology, The Ohio State University, 1925 Coffey Road, Columbus, OH 43210 USA and the [3]Ohio Environmental Protection Agency, 1800 Watermark Drive, Columbus, OH 43215, USA

Introduction

Many animals that form eggs with mineralized shells must increase calcium absorption from the intestinal tract and mobilize calcium from bone to provide a source of calcium for the egg shells and maintain normal concentrations of ionized calcium in the serum (Rosol *et al.* 1995). Stimulation of renal 1,25-dihydroxyvitamin D production by steroid hormones or parathyroid hormone, and stimulation of osteoclastic bone resorption by parathyroid hormone are mechanisms for maintaining serum calcium concentrations in the normal range. Morphologic studies on the parathyroid glands of freshwater turtles have not demonstrated seasonal variations (Srivastav *et al.* 1995). The aims of this investigation were to examine the gross, histologic, and ultrastructural morphology of parathyroid glands from female snapping turtles during egg-laying; male animals served as controls.

Experimental design and methods

Parathyroid glands were evaluated from adult snapping turtles (*Chelydra serpentina*; 23 females and 7 males) collected from the rivers of Ohio during June and July 1997. Serum total calcium, phosphorus, and albumin concentrations were measured in 33 females and 26 males. One or more parathyroid glands were fixed in 3% glutaraldehyde, post-fixed in 0.1 M cacodylate buffer and osmium, embedded in medcast plastic, and examined with a Philips 300 electron microscope or were fixed in 10% neutral-buffered formalin, embedded in paraffin, and examined by light microscopy. The numerical data are presented as means ± standard deviation and were evaluated by Student's unpaired *t*-test.

Results

The parathyroid glands were pink to yellow and two were found near the thymus and two were located near the aortic arches. The glands were larger and more easily identified in the females (Figs 1 and 2). The parathyroid glands (from the same anatomic location) from female turtles were two- to threefold larger in weight and

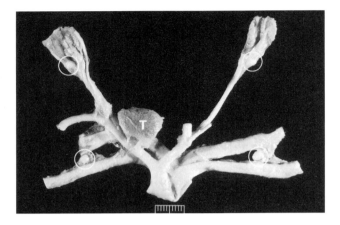

Fig. 1 Four hypertrophied parathyroid glands (circled) from a female snapping turtle with intra-abdominal eggs. Two glands are near the thymus (top) and two are near the aortic arches (bottom). The thyroid gland (T) is in the center. Scale: mm divisions.

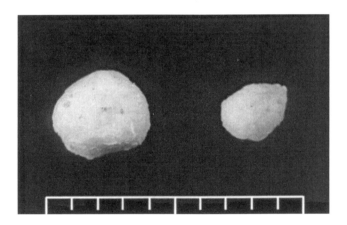

Fig. 2 Posterior (aortic) parathyroid glands from a female snapping turtle with intra-abdominal eggs (left; gland weight 13 mg, body weight 10.8 kg) and male adult snapping turtle (right; gland weight 4 mg, body weight 10.0 kg). Scale: mm divisions.

cross-sectional area than glands from male turtles. Random cross-sectional areas of parathyroid glands were 3.2 ± 2.2 mm^2 ($n=16$) in females and 1.1 ± 0.4 mm^2 ($n=4$) in males. Seventeen females collected during June had eggs ($n=6$-79 eggs; 73-1016 g total weight) with mineralized shells in their abdomens. Females collected during July ($n=16$) did not have eggs, but did have enlarged parathyroid glands. These females likely laid their eggs prior to collection of the turtles from the environment. All females (with or without eggs) had greater serum total calcium concentrations ($n=33$, 4.05 ± 0.88 mmol/l, range 3.08-5.85) compared with males ($n=26$, mean 2.18 ± 0.25 mmol/l, range 1.68-2.72) ($P<0.001$). The high serum calcium concentrations were considered normal due to the expected high serum vitellogen concentrations associated with egg-laying. Serum phosphorus concentrations were greater in the females (1.45 ± 0.39 mmol/l) compared with males (1.03 ± 0.29 mmol/l) ($P<0.001$). Serum albumin concentrations were mildly increased in females (18.0 ± 5.0 g/l) compared with males (14.7 ± 5.0 g/l) ($P<0.05$).

Histology and ultrastructure of parathyroid gland chief cells
Parathyroid glands consisted of cords of columnar chief cells. Most chief cells contained intracytoplasmic fat globules that resulted in the yellow coloration of glands grossly. Oxyphil cells were occasionally observed. Parathyroid chief cells from female turtles had ultrastructural evidence of increased cellular activity characterized by larger chief cells with prominent rough endoplasmic reticulum and Golgi apparatus, moderate numbers of mature secretory granules, the presence of prosecretory granules, and interdigitations of adjacent cell membranes compared with chief cells from male turtles.

Discussion
These data showed that parathyroid gland hypertrophy and hyperplasia occurred in normal female snapping turtles during egg-laying. In addition, the parathyroid gland chief cells had ultrastructural evidence of increased synthesis and secretion of parathyroid hormone. These changes likely represent a normal physiologic response to the increased demands for calcium mobilization during formation of egg shells. Interestingly, parathyroidectomy in fresh-water turtles (*Chrysemys picta, Pseudemys scripta, Graptemys pseudogeographica*) did not change the serum concentrations of calcium and phosphorus, but did increase phosphorus excretion (Clark 1965). This indicates that PTH may not play an important role in the daily regulation of serum calcium concentration in turtles. However, increased secretion of PTH may be necessary to maintain normal serum calcium concentrations during mineralization of egg shells. Parathyroid hormone can contribute to mineralization of egg shells by stimulating osteoclastic bone resorption or inducing increased synthesis of 1,25-dihydroxyvitamin D by the kidneys. Increased serum phosphorus concentrations in the female turtles of this study may have been due to the effects of 1,25-dihydroxyvitamin D on intestinal absorption of phosphorus or direct effects of parathyroid hormone on phosphorus excretion since parathyroidectomy increased phosphorus excretion in turtles.

References

Clark NB (1965) Experimental and histological studies of the parathyroid glands of fresh-water turtles. *General and Comparative Endocrinology* **5** 297-312.

Rosol TJ, Chew DJ, Nagode LA & Capen CC (1995) Pathophysiology of calcium metabolism. *Veterinary Clinical Pathology* **24** 49-63.

Srivastav AK, Sasayama Y & Suzuki N (1995) Morphology and physiological significance of parathyroid glands in reptilia. *Microscopy Research and Technique* **32** 91-103.

Part Four

Mammals

Mammalian calcium metabolism

T J Rosol

Department of Veterinary Biosciences, The Ohio State University, 1925 Coffey Road, Columbus, Ohio 43210, USA

Introduction
During evolution mammals developed in a terrestrial environment that was relatively low in abundance of calcium compared with seawater. Therefore, land mammals developed mechanisms to absorb and conserve calcium. Calcium homeostasis in mammals is principally regulated by parathyroid hormone (PTH) and coordinates its primary and secondary actions in bone, kidney and the intestinal tract with the active form of vitamin D (calcitriol). Synthesis of calcitriol in the kidneys is induced by PTH. Calcitriol is responsible for stimulating transcellular calcium transport in the intestines and completing an important negative feedback loop by inhibiting PTH synthesis. PTH-related protein (PTHrP) plays an important role in calcium homeostasis in the fetus where it regulates placental transport and serum concentration of calcium. Calcitonin is less important in mammals than in fishes and amphibians, and functions to limit postprandial rises in serum calcium concentration. Abnormal regulation of calcium balance occurs most commonly in mammals when there are dietary imbalances in calcium, phosphate, or vitamin D, and in animals with certain forms of cancer that induce humoral hypercalcemia of malignancy. Most diseases of calcium metabolism in humans have counterparts in non-human mammals, and these animals can be used to represent spontaneous *in vivo* models of human disease.

Calcium in the body
The majority of the calcium (Ca) of the body (99%) is present in the inorganic matrix of bone as hydroxyapatite. Most of the remaining Ca (0.9%) is associated with the plasma membrane and sequestered in the endoplasmic reticulum of cells. Extracellular fluid contains 0.1% of the body's Ca mass, with a total Ca concentration of about 2.5 mM. Approximately 55% of the extracellular Ca (1.4 mM) is in the ionized form (Ca^{2+}), the biologically active form of Ca. Neonatal animals have slightly higher extracellular Ca concentrations than adults. There is very little ionized Ca in the cytosol of cells (approximately 100 nM).

Forms of Ca
Extracellular and serum Ca exists in three forms: (1) ionized Ca^{2+} (55%); (2) Ca^{2+} (10%) complexed to anions; (3) protein-bound Ca^{2+} (35% of total Ca) (Rosol *et al.* 1999). The protein-bound fraction of Ca^{2+} is principally bound to negatively charged sites on albumin. The protein-bound form of Ca^{2+} is dependent on serum pH, and

alterations in serum pH will change the [Ca^{2+}]. As the pH of serum becomes more acidic, [Ca^{2+}] increases because of competition with H^+ for binding to the negatively charged sites on serum proteins.

Function of Ca

Ca serves two primary functions in the body: (1) structural in bones and teeth; (2) as a messenger or regulatory ion. The 10 000-fold concentration gradient of ionized Ca^{2+} between the extracellular fluid (1.4 mM) and the cytoplasm (100 nM) permits Ca^{2+} to function as a signaling ion to activate intracellular processes. The lipid bilayer of the cell membrane has a low permeability to Ca^{2+}. Therefore, influx of Ca^{2+} into the cytoplasm is controlled by voltage-operated or receptor-operated Ca channels. Influx of Ca^{2+} into cells can (1) regulate cellular function by interactions with Ca-binding proteins (e.g. calmodulin) and (2) stimulate biological responses such as neurotransmitter release, muscle contraction, and secretion. Ca also plays an important role in fertilization, development, cell division, cell adhesion, and blood coagulation. The responses of cells to a rise in intracellular Ca^{2+} can vary depending on the spatial aspects of the rise (local vs global), the oscillation frequency of the rise (0.1 s to hours), and the amplitude of the rise (Berridge *et al.* 1998). Ionized Ca also may regulate cellular function by binding to Ca^{2+}-specific receptors in the cell membrane, such as in parathyroid chief cells or renal epithelial cells (Pollak *et al.* 1993). It is critical for cells to maintain the normal low level of intracellular Ca^{2+}. If cellular Ca homeostasis fails because of anoxia, an energy-deprived state, or perturbed membrane integrity, cell viability is threatened because of uncontrolled entry of Ca^{2+}, which can lead to cell death and apoptosis (Berridge *et al.* 1998).

Measurement of Ca

Total serum or urine Ca concentrations can be measured directly using atomic absorption spectrophotometry or a colorimetric assay using an *o*-cresolphthalein complex (Fraser *et al.* 1987). Animals with hypoalbuminemia or hypoproteinemia will have decreased total serum Ca concentrations because of a reduction in protein-bound Ca. Blood or plasma [Ca^{2+}] can be measured using an instrument with a Ca^{2+}-selective electrode (Fraser *et al.* 1987). It is most informative to measure the biologically active form of Ca (ionized Ca^{2+}), but this is not always practical because of the need for special instrumentation and proper sample handling. Anaerobic collection is critical to prevent changes in the pH of the sample (Schenck *et al.* 1995).

Daily Ca balance and movement

Ca is absorbed in the gastrointestinal tract, with greatest absorption occurring in the duodenum (McDowell 1992). In normal animals an equivalent amount of Ca is excreted primarily in the urine with small losses in sweat and intestinal secretions. The Ca released and deposited daily from bone are balanced in adult animals. Therefore, intestinal Ca absorption is the major determinant of the amount of Ca excreted in the urine in adults. Dietary sources of Ca for mammals under natural conditions are usually adequate and are not limiting unless food supply in general is scarce

(McDowell 1992). Carnivores, omnivores, and insectivores obtain adequate Ca from ingestion of exo- and endoskeletons. Herbivores obtain Ca from grasses (0.31-0.36% Ca) and legumes (1.2-1.7% Ca). Seeds (grains) are low in Ca content (0.02-0.1%). Some dietary components impair intestinal Ca availability, such as oxalate or phytic acid. Diets not only must contain adequate available Ca, but also a proper ratio of calcium to phosphorus (approximately 1.5-2:1). Diets with excessive phosphate or a low Ca:P ratio will lead to impaired Ca absorption. Grains are high in phosphate and low in calcium, so diets with a predominance of seed grains must be formulated properly to maintain normal Ca balance.

Ca can move across epithelial barriers by two routes: (1) transport between cells (paracellular) and (2) transport through cells (Brown 1994). Transport of Ca between cells occurs by two mechanisms, convection and diffusion. Convection (solvent drag) is the movement of Ca with the flow of water. Convection of Ca occurs in the process of glomerular filtration and with water reabsorption in the proximal convoluted tubules. Diffusion is the passive movement of Ca down an electrical or chemical gradient.

In order for Ca to move through cells, mechanisms have evolved for the regulation of Ca transport into the cytosol, buffering in the cytoplasm, and release from the cell (Brown 1994). Large concentration and electrical gradients are present for Ca^{2+} to enter cells. There is a 10 000-fold concentration gradient from the extracellular to intracellular fluids and the interior of cells are negatively charged (-50 to -100 mV). After cell membrane Ca^{2+} channels are opened, there is rapid diffusion of Ca^{2+} into the cell. The rise in intracellular Ca^{2+} can stimulate biological processes by interaction with Ca^{2+}-binding enzymes and regulatory proteins (Berridge et al. 1998). The Ca^{2+} that enters the cytosol is rapidly buffered by Ca-binding proteins and reduced by transport into the endoplasmic reticulum or extracellularly by Ca^{2+}-ATPases or Na^+/Ca^{2+} exchangers.

Renal handling of calcium

The kidney normally reabsorbs 98% or more of the filtered Ca. This high degree of reabsorption is important to maintain the balance of Ca in the body but permits the kidneys, if necessary, to excrete large amounts of Ca in the urine (Bindels 1993). Most filterable Ca (ionized and complexed) enters the glomerular filtrate by convection and is reabsorbed by the renal tubules. The kidneys reabsorb approximately 40-fold more Ca than is absorbed by the intestine because of the high degree of blood flow and ultrafiltration in the glomerulus.

Most filtered Ca (70%) is reabsorbed in an unregulated manner in the proximal convoluted tubules by diffusion and convection with water uptake between the epithelial cells. The thick ascending loop of Henle absorbs about 20% of the filtered Ca, but the precise mechanism is unclear. The site of active regulation of Ca reabsorption is the distal convoluted tubule, which reabsorbs approximately 10% of the filtered Ca. The principal stimulator of Ca reabsorption in the distal convoluted tubule is PTH.

The serum Ca^{2+} concentration can directly regulate Ca and water excretion by intrinsic responses mediated by Ca^{2+}-sensing receptors on the renal epithelial cells (Brown & Hebert 1997). Decreased urinary concentrating ability and polyuria are early functional effects of increased serum Ca^{2+} and results from a reduced reabsorption of sodium and impaired action of antidiuretic hormone. Additional direct effects of serum Ca^{2+} on the kidney include inhibition of tubular Ca reabsorption and antagonism to the actions of PTH. These responses by the kidney facilitate Ca excretion and help to ameliorate the clinical effects of hypercalcemia in animals.

Intestinal Ca absorption

There are two components of Ca absorption from the intestinal tract, namely saturable or transcellular transport and non-saturable or intercellular (paracellular) transport (Favus 1992). Low-Ca diets are associated with high absorption rates (up to 95%), and high-Ca diets have low absorption rates (about 40%). Although high-Ca diets have low absorption rates, they can increase serum Ca concentrations by non-saturable intestinal absorption. In contrast, diets deficient in Ca are associated with normal serum Ca concentrations because of compensation by PTH-stimulated bone resorption, renal Ca reabsorption, and increased calcitriol synthesis.

Saturable transport is a carrier-mediated vitamin D-dependent process and occurs predominantly in the duodenal segment of the small intestine, but also occurs in the cecum and colon (Favus 1992, Karbach & Feldmeier 1993). The active form of vitamin D (1,25-dihydroxyvitamin D, calcitriol) stimulates transcellular transport of Ca^{2+}. One function of calcitriol in the intestinal epithelial cell is to increase the expression of calbindin, an intracellular Ca-binding protein. Non-saturable Ca transport occurs throughout the small intestine and is the main mechanism for Ca absorption in animals with vitamin D deficiency. Non-saturable Ca^{2+} transport is dependent on the luminal $[Ca^{2+}]$. As dietary intake of Ca increases, much of the Ca in the intestinal lumen is unavailable for absorption because of precipitation of Ca salts or complexes formed with anions (McDowell 1992).

Absorption of Ca is increased during pregnancy, lactation, growth, and when animals are fed low-Ca diets. The primary adaptive influence on Ca absorption is the circulating calcitriol concentration. Factors that increase intestinal Ca absorption directly or indirectly by stimulation of calcitriol synthesis include PTH, growth hormone, testosterone, and estrogen. Factors that reduce intestinal absorption of Ca include glucocorticoids, thyroid hormones, chronic acidosis, and luminal conditions (phosphate, phytates, oxalate, fatty acids, pH > 6.1, and other anions) that induce complexation of Ca^{2+} (Favus 1992).

Bone and Ca balance

There are two sources of Ca^{2+} in bone that can enter the circulation: (1) readily mobilizable Ca salts in the extracellular fluid; (2) hydroxyapatite crystals which require 'digestion' by osteoclasts before Ca can be released from bone. The nature and regulation of the readily mobilizable Ca in bone is poorly understood; however, it is present in small amounts and probably plays a role in the fine regulation of serum Ca

Calcium Metabolism: Comparative Endocrinology

concentration (Parfitt 1987). If there is a significant need for Ca from bone, it must come from osteoclastic resorption of hydroxyapatite crystals. In adult animals there is a stable balance between Ca deposition associated with bone formation and Ca release associated with osteoclastic bone resorption. In young animals, bone has a positive Ca balance because of the relative excess of bone formation. Conditions that result in excessive bone resorption (e.g. cancer-associated hypercalcemia or primary hyperparathyroidism) can cause the release of large amounts of Ca from bone and contribute to the development of hypercalcemia (Rosol & Capen 1992).

Parathyroid hormone
PTH is an 84 amino acid peptide produced by the chief cells of the parathyroid glands. The circulating concentration of PTH depends on regulation of PTH gene transcription, intracellular catabolism, and secretion (Kronenberg *et al.* 1994). Under normal conditions much of the PTH produced is degraded intracellularly; therefore, PTH production by the chief cells can be rapidly increased by reducing the amount of catabolism. The major stimulus for increased PTH synthesis and secretion is a reduction in serum Ca^{2+}.

Serum Ca^{2+} binds to a transmembrane receptor on the chief cell which permits the serum $[Ca^{2+}]$ to regulate chief cell function. Interaction of serum Ca^{2+} with its receptor on chief cells results in an inverse sigmoidal relationship between serum Ca^{2+} and PTH concentrations. The serum $[Ca^{2+}]$ that results in half-maximal PTH secretion is defined as the serum Ca 'set point' and is stable for an individual animal. The sigmoidal relationship between serum $[Ca^{2+}]$ and PTH secretion permits the chief cells to respond rapidly to a reduction in serum Ca^{2+}. Mutations in one or both of the Ca^{2+}-sensing receptor genes in humans results in familial hypocalciuric hypercalcemia or neonatal severe hypercalcemia, respectively (Pollak *et al.* 1993).

The major inhibitors of PTH synthesis and secretion are increased serum $[Ca^{2+}]$ and calcitriol. Inhibition of PTH synthesis by calcitriol completes an important endocrine feedback loop between the parathyroid chief cells and the renal epithelial cells as PTH stimulates renal production of calcitriol. In addition, serum phosphate has been identified as a regulator of PTH mRNA stability (Moallem *et al.* 1998). High serum phosphate concentrations will increase PTH synthesis and secretion.

PTH has two functional domains. N-Terminal PTH (PTH(1-34)) is the region of major biological activity in relation to Ca regulation. The function of C-terminal PTH peptides is less well understood (Mallette 1994). PTH is secreted in two forms: intact PTH(1-84) and C-terminal peptides. PTH(1-84) is the circulating form of biologically active PTH. It is rapidly metabolized by endopeptidases on Kupffer cells in the liver (Arnaud & Pun 1992). Intact serum PTH concentrations are best measured by two-site immunoradiometric assay (IRMA; Nussbaum & Potts 1994).

The receptor for the N-terminal region of PTH is a seven-transmembrane domain receptor that is expressed in target cells including renal epithelial cells and osteoblasts (Segre 1994). Binding of PTH to its receptor results in increased cytoplasmic cAMP and Ca^{2+} as the result of stimulation of adenylyl cyclase and phospholipase C (Coleman *et al.* 1994).

The primary function of intact PTH is to maintain normal serum Ca^{2+} concentration by the stimulation of renal Ca reabsorption in the distal convoluted tubules, increased release of readily mobilizable Ca from bone, increased osteoclastic bone resorption, and increased gastrointestinal absorption of Ca^{2+}. The increase in Ca^{2+} absorption from the gastrointestinal tract is due principally to indirect stimulation of renal calcitriol production. PTH is responsible for the fine minute-to-minute regulation of serum $[Ca^{2+}]$ predominantly by regulating renal Ca reabsorption. Stimulation of renal calcitriol production and osteoclastic bone resorption requires more time (hours to days), but has the potential to provide larger amounts of Ca.

1,25-Dihydroxyvitamin D (calcitriol)

The active form of vitamin D, calcitriol, is produced by 1α-hydroxylation of 25-hydroxyvitamin D by epithelial cells of the proximal convoluted tubules in the kidney. This enzymatic reaction is the rate-limiting step in the formation of calcitriol. The primary functions of calcitriol in Ca regulation include stimulation of intestinal Ca absorption, inhibition of PTH synthesis by decreasing PTH mRNA transcription, promoting osteoclastic bone resorption, and negative feedback on its own synthesis in renal epithelial cells.

Renal production of calcitriol is increased by PTH and hypophosphatemia. Renal synthesis of calcitriol is increased by growth hormone, estrogen, and prolactin, which are important hormones for maintaining Ca balance during growth, pregnancy, and lactation, respectively. Renal production of calcitriol is inhibited by hyperphosphatemia, hypercalcemia, and renal diseases characterized by loss of renal tubular mass. Since the structure of calcitriol does not vary between species, the same assays can be used to measure serum 1,25-dihydroxyvitamin D in animals and humans. Horses have very low circulating concentrations of calcitriol (<5 pg/ml) which suggests that intestinal calcium absorption is less dependent on vitamin D in this species.

Calcitonin

Calcitonin is a 32 amino acid peptide that functions to reduce serum $[Ca^{2+}]$. It is not a major factor in the minute-to-minute regulation of serum Ca^{2+}, but serves as an 'emergency' hormone to reduce serum Ca^{2+} at times of rapid increases in serum Ca^{2+} (Rosol & Capen 1997). It is stored in high levels in the cytoplasm of thyroid C-cells so that it can respond quickly to increased serum Ca^{2+} concentrations. The secretion of calcitonin is stimulated by hypercalcemia and ingestion of high-Ca meals by enteric secretion of gastrin and cholecystokinin. The primary function of calcitonin is to inhibit osteoclastic bone resorption; however, its effect on osteoclasts is transitory. Osteoclasts soon become refractory to the effects of calcitonin, which limits its effectiveness in controlling acute increases in serum Ca^{2+}. Chronic hypercalcemia results in prolonged increases in serum calcitonin concentrations and C-cell hyperplasia with little biological significance in relation to serum Ca regulation.

Calcium Metabolism: Comparative Endocrinology

Table 1 Important diseases of abnormal Ca metabolism in animals (with examples)

Disorders associated with normal serum Ca concentration

Rickets (vitamin D or phosphorus deficiency in growing animals)
- Pigs housed indoors or fed unbalanced diets

Rickets (hereditary defects in vitamin D action)
- Type II: target organ resistance to calcitriol – New World primates (common marmoset)

Osteomalacia (vitamin D or phosphorus deficiency in adult animals)

Osteoporosis
- Animals fed inadequate amounts of feed
- Herbivores fed poor quality and inadequate roughage (winter grazing)

Nutritional secondary hyperparathyroidism (low Ca/high P diets)
- Carnivores (dogs, cats) fed all meat or viscera diets
- Monkeys fed unsupplemented diets
- Horses, herbivores, and pigs fed unsupplemented high-grain or bran diets
- Metabolic bone disease (iguanas, due to lack of UV light and dietary Ca)
- Primary Ca deficiency (rare)

Renal secondary hyperparathyroidism
- Animals with chronic renal disease

Disorders associated with hypercalcemia

Humoral hypercalcemia of malignancy
- Dogs with lymphoma, apocrine adenocarcinoma of the anal sac, or miscellaneous carcinomas
- Cats with lymphoma or miscellaneous carcinomas
- Horses with squamous cell carcinoma

Hypervitaminosis D
- Ingestion of plants with calcitriol glycosides
- Ingestion of rodenticides with cholecalciferol

Local osteolysis
- Osteomyelitis
- Multiple myeloma (dogs, cats)
- Metastatic carcinoma (rare)

Hypervitaminosis A (cats, pigs, calves)

Primary hyperparathyroidism (parathyroid gland adenoma or hyperplasia)

Table 1 continued

	Dogs and cats (uncommon), horses (rare)
Renal secondary hyperparathyroidism	
	Horses with chronic renal failure (uncommon in other species)
Idiopathic hypercalcemia in cats	
Hypoadrenocorticism	
	Dogs with immune-mediated adrenalitis
Granulomatous diseases	
	Blastomycosis in dogs
Familial hypocalciuric hypercalcemia/neonatal severe hyperparathyroidism (mutation in the cell membrane Ca^{2+} receptor)	
	German Shepherd dogs (suspected)
Hypocalcemic disorders	
Postparturient paresis	
	High-producing dairy cows
	Small breed dogs
Preparturient paresis	
	Sheep or goats with multiple offspring
Hypoparathyroidism	
	Immune-mediated parathyroiditis in dogs
Acute or chronic renal failure (uncommon)	
Rickets (hereditary defects in vitamin D action)	
	Type I: Defect in calcitriol synthesis – pigs, dogs (rare), cats (rare)
Hypomagnesemia	
	Cattle with grass tetany
Stress-associated hypocalcemia	
	Prolonged transport without adequate food and hydration
	Endurance exercise trials

For a description of the above conditions, refer to the following references: Palmer (1993), Rosol & Capen (1997) and Rosol *et al.* (1999).

The primary structure of calcitonin varies considerably between species. This has two important consequences. First, the biological activity of calcitonin varies between species. Fish calcitonin (e.g. salmon calcitonin) has a 25-fold higher activity in mammals, which permits salmon calcitonin to be used as an effective pharmaceutical

Calcium Metabolism: Comparative Endocrinology

agent. Secondly, antibodies to calcitonin have poor cross-reactivity between species and radioimmunoassays for calcitonin are highly species-specific.

Parathyroid hormone-related protein
PTHrP is a 139-173 amino acid peptide originally isolated from human and animal tumors associated with humoral hypercalcemia of malignancy (Grill & Martin 1994, Rosol & Capen 1992). PTHrP shares 70% sequence homology with PTH in its first 13 amino acids. The N-terminal region of PTHrP (amino acids 1-34) binds and stimulates PTH receptors in bone and kidney cells with equal affinity to that of PTH with the result that PTHrP functions similarly to PTH *in vivo*.

PTHrP has been demonstrated in many normal tissues including epithelial cells of the skin and other organs, endocrine glands, smooth, skeletal, and cardiac muscle, lactating mammary gland, placenta, fetal parathyroid glands, bone, brain, and lymphocytes (Philbrick *et al.* 1996). The function of PTHrP in most tissues is poorly defined, but it probably functions as an autocrine or paracrine factor, as circulating concentrations of PTHrP in normal humans and animals are low (<1 pM; Rosol *et al.* 1992). It is not known if there is a specific receptor for PTHrP; however, the mid- and C-terminal regions of PTHrP have functions that are independent of its PTH-like effects (Mallette 1994). The midregion of PTHrP is responsible for stimulating Ca uptake by the fetal placenta (Abbas *et al.* 1989) and the C-terminal region can inhibit osteoclastic bone resorption (Fenton *et al.* 1991).

Fetuses maintain higher concentrations of serum Ca than their dams. As fetal parathyroid glands produce low levels of PTH, the mechanism of maintaining increased serum Ca concentrations in fetuses was not known until recently. Investigations have demonstrated that PTHrP functions to maintain Ca balance in the fetus (MacIsaac *et al.* 1991). It is the major hormone secreted by fetal parathyroid chief cells and is produced by the placenta to stimulate Ca uptake by the fetus. PTHrP also plays a role in differentiation of fetal tissues, especially in bone. Disruption of the PTHrP gene in transgenic mice leads to death at birth as the result of premature ossification of the skeleton, shortened bones, and respiratory failure because the thoracic cavity is unable to expand (Karaplis *et al.* 1994).

Biologically active PTHrP is secreted from certain malignant tumors in animals (especially dogs) and results in the syndrome humoral hypercalcemic malignancy by increasing circulating concentrations of PTHrP (Rosol & Capen 1992). Two-site immunoassays are available for the measurement of human PTHrP. These assays are useful for measuring PTHrP in animals (Rosol & Capen 1992) because of the high degree of sequence homology in PTHrP between species, especially in the first 111 amino acids (Rosol *et al.* 1995).

Important diseases of abnormal calcium metabolism in animals
Disorders of calcium metabolism are common in domestic mammals (Palmer 1993, Rosol & Capen 1997, Rosol *et al.* 1999). These diseases are important because of their effects on animal husbandry and their potential to serve as models of human disease. Table 1 gives a list of important diseases and examples.

References

Abbas SK, Pickard DW, Rodda CP, Heath JA, Hammonds RG, Wood WI, et al. (1989) Stimulation of ovine placental calcium transport by purified natural and recombinant parathyroid hormone-related protein (PTHrP) preparations. *Quarterly Journal of Experimental Physiology* **74** 549-552.

Arnaud CD & Pun K-K (1992) Metabolism and assay of parathyroid hormone. *Disorders of Bone and Mineral Metabolism*, pp 107-122. Eds FL Coe and MJ Favus. New York: Raven Press.

Berridge MJ, Bootman MD & Lipp P (1998) Calcium: a life and death signal. *Nature* **395** 645-648.

Bindels RJM (1993) Calcium handling by the mammalian kidney. *Journal of Experimental Biology* **184** 89-104.

Brown EM (1994) Homeostatic mechanisms regulating extracellular and intracellular calcium metabolism. In *The Parathyroids*, pp 15-54. Eds JP Bilezikian, R Marcus & MA Levine. New York: Raven Press.

Brown EM & Hebert SC (1997) Calcium receptor-regulated parathyroid and renal function. *Bone* **20** 303-309.

Coleman DT, Fitzpatrick LA & Bilezikian JP (1994) Biochemical mechanisms of parathyroid hormone action. In *The Parathyroids*, pp 239-258. Eds JP Bilezikian, R Marcus & MA Levine. New York: Raven Press.

Favus MJ (1992) Intestinal absorption of calcium, magnesium, and phosphorus. In *Disorders of Bone and Mineral Metabolism*, pp 57-81. Eds FL Coe & MJ Favus. New York: Raven Press.

Fenton AJ, Kemp BE, Kent GN, Moseley JM, Zheng M-H, Rowe DJ, et al. (1991) A carboxyl-terminal peptide from the parathyroid hormone-related protein inhibits bone resorption by osteoclasts. *Endocrinology* **129** 1762-1768.

Fraser D, Jones G, Kooh SW & Radde IC (1987) Calcium and phosphate metabolism. In *Fundamentals of Clinical Chemistry*, pp 705-728. Ed NW Tietz. Philadelphia: WB Saunders.

Grill V & Martin TJ (1994) Parathyroid hormone-related protein as a cause of hypercalcemia in malignancy. In *The Parathyroids*, pp 295-311. Eds JP Bilezikian, R Marcus & MA Levine. New York: Raven Press.

Karaplis AC, Luz A, Glowacki J, Bronson RT, Tybulewicz VLJ, Kronenberg HM, et al. (1994) Lethal skeletal dysplasia from targeted disruption of the parathyroid hormone-related peptide gene. *Genes and Development* **8** 277-289.

Karbach U & Feldmeier H (1993) The cecum is the site with the highest calcium absorption in rat intestine. *Digestive Diseases and Sciences* **38** 1815-1824.

Kronenberg HM, Bringhurst FR, Segre GV & Potts Jr JT (1994) Parathyroid hormone biosynthesis and metabolism. In *The Parathyroids*, pp 125-138. Eds JP Bilezikian, R Marcus & MA Levine. New York: Raven Press.

MacIsaac RJ, Heath JA, Rodda CP, Moseley JM, Care AD, Martin TJ, et al. (1991) Role of the fetal parathyroid glands and parathyroid hormone-related protein in the regulation of placental transport of calcium, magnesium, and inorganic phosphate. *Reproduction, Fertility and Development* **3** 447-457.

Mallette LE (1994) Parathyroid hormone and parathyroid hormone-related protein as polyhormones. In *The Parathyroids*, pp 171-184. Eds JP Bilezikian, R Marcus & MA Levine. New York: Raven Press.

McDowell LR (1992) Calcium and phosphorus. *Minerals in Animal and Human Nutrition,* pp 26-77. San Diego: Academic Press.

Moallem E, Kilav R, Silver J & Naveh-Many T (1998) RNA-protein binding and post-transcriptional regulation of parathyroid hormone gene expression by calcium and phosphate. *Journal of Biological Chemistry* **273** 5253-5259.

Nussbaum SR & Potts JT Jr (1994) Advances in immunoassays for parathyroid hormone. In *The Parathyroids,* pp 157-170. Eds JP Bilezikian, R Marcus & MA Levine. New York: Raven Press.

Palmer N (1993) Bones and joints. *Pathology of Domestic Animals,* pp 1-182. Eds KVF Jubb, PC Kennedy & N Palmer. San Diego: Academic Press.

Parfitt AM (1987) Bone and plasma calcium homeostasis. *Bone* **8** (Suppl.) 1 S1-S8.

Philbrick WM, Wysolmerski JJ, Galbrath S, Holt E, Orloff JJ, Yang KH, *et al.* (1996) Defining the roles of parathyroid hormone-related protein in normal physiology. *Physiology Reviews* **76** 127-173.

Pollak MR, Brown EM, Chou Y-HW, Hebert SC, Marx SJ, Steinmann B, *et al.* (1993) Mutations in the human Ca^{2+}-sensing receptor gene cause familial hypocalciuric hypercalcemia and neonatal severe hypercalcemia. *Cell* **75** 1297-1303.

Rosol TJ & Capen CC (1992) Biology of disease: mechanisms of cancer-induced hypercalcemia. *Laboratory Investigations* **67** 680-702.

Rosol TJ & Capen CC (1997) Calcium-regulating hormones and diseases of abnormal mineral (calcium, phosphorus, magnesium) metabolism. In *Clinical Biochemistry of Domestic Animals,* pp 619-702. Eds JJ Kaneko, JW Harvey & ML Bruss. San Diego: Academic Press.

Rosol TJ, Nagode LA, Couto CG, Hammer AS, Chew DJ, Peterson JL, *et al.* (1992) Parathyroid hormone (PTH)-related protein, PTH, and 1,25-dihydroxyvitamin D in dogs with cancer-associated hypercalcemia. *Endocrinology* **131** 1157-1164.

Rosol TJ, Steinmeyer CL, McCauley LK, Gröne A, DeWille JW & Capen CC (1995) Sequences of the cDNAs encoding canine parathyroid hormone-related protein and parathyroid hormone. *Gene* **160** 241-243.

Rosol TJ, Chew DJ, Nagode LA & Schenck PA (1999) Disorders of calcium: hypercalcemia and hypocalcemia. In *Fluid Therapy in Small Animal Practice*. Ed SP DiBartola. Philadelphia: WB Saunders Co (In Press).

Schenck PA, Chew DJ & Brooks CL (1995) Effects of storage on normal canine serum ionized calcium and pH. *American Journal of Veterinary Research* **56** 304-307.

Segre GV (1994) Receptors for parathyroid hormone and parathyroid hormone-related protein. In *The Parathyroids,* pp 213-230. Eds JP Bilezikian, R Marcus & MA Levine. New York: Raven Press.

Parathyroid hormone-related peptide may play a role in deer antler regeneration

C Faucheux and J S Price

The Bone and Mineral Centre, University College London, The Rayne Institute, 5 University Street, London WC1E 6JJ, UK

Introduction

The deer antler is the only mammalian organ capable of complete regeneration, which happens annually after casting of the previous year's growth. Antlers grow rapidly from a mass of undifferentiated cells that fits the definition of a blastema, by a process of modified endochondral ossification (Price et al. 1996), until rising levels of testosterone induce rapid mineralisation and slow longitudinal growth (Goss 1983). The endocrinology of antler growth has been well documented. However, to date, it remains unclear how systemic hormones interact with paracrine and autocrine factors to regulate cellular proliferation and differentiation in the antler.

A molecule known to play an important role in regulating bone development in other systems is parathyroid hormone-related peptide (PTHrP). Endochondral ossification is abnormal in mice that lack or overexpress the PTHrP gene, and overexpression of the PTH/PTHrP receptor leads to a form of human chondrodysplasia (Karapalis et al. 1994, Schipani et al. 1995, Weir et al. 1996). The aim of this study was to explore the hypothesis that PTHrP also functions as an autocrine/paracrine factor during mammalian bone regeneration.

Materials and methods

Animals
Eight red deer stags (~2 years old) were killed for venison and their antlers harvested *post mortem* at different time points after the previous set had been cast (3 days to 6 weeks).

Cell culture
The different regions of the antler tip are illustrated in Fig. 1 and cell digests were prepared from the skin (velvet), perichondrium, zone of chondroprogenitors and unmineralised cartilage as previously described (Price et al. 1994). Cells were cultured in growth medium (GM; BGJb medium; GibcoBRL, Glagow, Scotland, supplemented with 10% fetal calf serum and antibiotics), subcultured when confluent and used at first passage unless otherwise stated. Primary cultures of cells from the cartilage region were also cultured as a micromass.

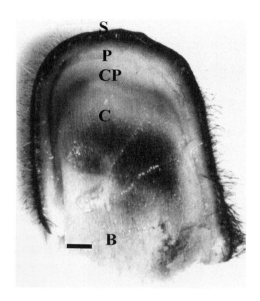

Fig. 1 Longitudinal section through the growing antler tip. S, velvet skin; P, perichondrium; CP, chondroprogenitors; C, non-mineralised cartilage; B, spongy bone. Bar=500 μm.

Reverse transcriptase-polymerase chain reaction (RT-PCR)
PTH and PTH/PTHrP receptor mRNA expression was studied in tissue harvested from the antler blastema (4 days of growth) and from rapidly growing antlers (3-4 weeks) where the different regions of the distal tip are distinguishable macroscopically. A 2 μg portion of extracted total RNA (Ultraspec RNA; Biogenesis, Poole, UK) was reverse transcribed and RT-PCR analysis undertaken, using primers specific for human PTHrP (5'-ACCTCGGAGGTGTCCCCTAAC-3'; 5'-CGTGAATCGAGCTCCAGCGACG-3') and PTH/PTHrP receptor (5'-CTGGTGGATGCAGATGACGTC-3'; 5'-CAGCTGCCATTGCGGTCACAG-3'.

RT-PCR cycling conditions were as follows: 94°C for 4 min, 35 cycles of 94°C for 1 min, 67°C for 1.5 min and 72°C for 1 min, then 72°C for 10 min. The identity of the PCR products was confirmed by sequencing using a genetic analyser (Perkin-Elmer, Norwalk, CT, USA).

Synthesis of PTHrP
Primary cultures from skin, perichondrium and cartilage were seeded into 12-well plates (10^5 cells/well) and cultured in GM until confluent (3-4 days). PTHrP levels in the medium were measured using a human IRMA (Mitsubishi Chemical Co, Tokyo, Japan). The species cross-reactivity of the assay was assessed by parallelism of a serial

dilution of samples from antler cells with the standard curve. The amount of protein per well served as reference.

[^{3}H]Thymidine incorporation

Cells were seeded on to 48-well plates (10^4 cells/well) and cultured in GM for 24 h. Cells were deprived of serum for 24 h in BGJb medium supplemented with 0.1% fetal calf serum, then incubated for 24 h in the presence or absence of PTHrP(1-34) (10^{-7}-10^{-9} M; Bachem, Saffron Walden, Essex, UK) in medium containing 2% fetal calf serum and [^{3}H]thymidine (1 µCi/well). [^{3}H]Thymidine incorporation was determined as previously described (Price et al. 1994).

Alkaline phosphatase (ALP) assay

Cells were seeded on to 12-well plates (5×10^4 cells/well) and maintained in GM until confluent. For micromass experiments, a 10 µl (2×10^5 cells) suspension of cells digested from cartilage was spotted on to the surface of 12-well plates. Cultures were incubated with PTHrP (10^{-7} M) for 5 days. ALP activity in cell lysates was assayed as previously described (Price et al. 1994) and expressed as µmol/h per µg protein.

Statistical analysis

All in vitro experiments were repeated three times unless otherwise stated (n=3-4 wells). Results of representative experiments are presented. Data were analysed with a Statview 512 software package, and statistical significance of differences between the means of test conditions and respective controls was assessed by analysis of variance (post hoc examination with Scheffé's test).

Results

PTHrP and PTH/PTHrP receptor mRNA expression

PTHrP mRNA was detected in tissues from the blastema and at later stages of growth in the skin, perichondrium and zone of chondroprogenitors. However, repetitive RT-PCR showed a weak and infrequent signal in cartilage (Fig. 2A). For the PTH/PTHrP receptor, a strong signal was observed in the blastema, the perichondrium and the non-mineralised cartilage but not in skin (Fig. 2B). Both PTHrP and PTH/PTHrP receptor mRNAs were expressed in cells cultured from all regions. Sequence analysis of PCR products revealed that cervid PTHrP and PTH/PTHrP receptor had ~89% identity with human nucleotide sequences.

Synthesis of PTHrP by cultured cells

When conditioned medium was serially diluted, the curve drawn was parallel to the assay standard curve, demonstrating species cross-reactivity. PTHrP was detected in medium conditioned by primary cells. However, there were significant differences in the level of synthesis between cell types: skin>perichondrium>cartilage (Fig. 3).

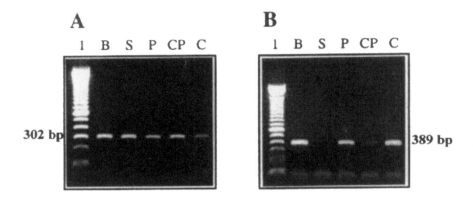

Fig. 2 Expression of PTHrP (A) and PTH/PTHrP receptor (B) mRNA in deer antler. Total RNA was extracted from tissues of two stages of growth: early stage (B, blastema) and middle stage (S, skin; P, perichondrium; CP, chondroprogenitor region; C, non-mineralised cartilage). Lane 1, 100 bp DNA ladder.

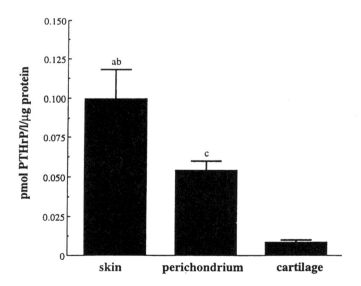

Fig. 3 Synthesis of secreted PTHrP. PTHrP was measured in medium conditioned by primary cultures of cells derived from skin, perichondrium and cartilage. PTHrP production is expressed as pmol/l per µg protein. Mean±S.D. ($n=3$). a: $P<0.001$ relative to perichondrium; b: $P<0.0001$ relative to cartilage; c: $P<0.001$ relative to cartilage.

Effect of PTHrP on [^{3}H]thymidine incorporation

PTHrP had no effect on the proliferation of cells cultured from the antler blastema or from the perichondrium. However, PTHrP treatment (10^{-7}-10^{-9}M) increased [^{3}H]thymidine incorporation in cells cultured from cartilage (Table 1). The proliferative response to PTHrP appeared to be dependent on the stage of cellular differentiation; it was greater in cells derived from antlers at a late stage of growth which had higher basal levels of ALP activity.

Table 1 Stimulation of [^{3}H]thymidine incorporation in deer antler cartilage derived cells.

Deer	Percentage of stimulation			ALP activity (nmolePi/h/µg protein)
	PTHrP (10^{-9} M)	PTHrP (10^{-8} M)	PTHrP (10^{-7} M)	
Deer 2 (10 days)	ND	120**	120**	10±1
Deer 7 (3 weeks)	110*	113**	115**	24±5
Deer 5 (9 weeks)	150**	135*	190***	259±5

Values are expressed as percentage increase compared to control values. Alkaline phosphatase (ALP) activity is expressed as mean±S.D. Time of growth is indicated in brackets.
ND: non determined. Pi: inorganic phosphate. *$P<0.05$, **$P<0.001$, ***$P<0.0001$

Effect of PTHrP on ALP activity

PTHrP had no effect on ALP activity in cells cultured from the blastema or the perichondrium which express low basal levels of ALP (0.001-0.014 µmol/h per µg protein). Similarly PTHrP had no effect on ALP activity in cartilage cells cultured as a monolayer from the same antler. However, when cartilage-derived cells were cultured as a micromass, PTHrP treatment significantly decreased ALP activity (Fig. 4). These culture conditions are more appropriate for maintaining the cell's phenotype; they express a higher level of ALP activity and there is positive alcian blue staining of the micromass (data not shown).

Discussion

PTHrP was originally identified as a factor that mediates the humoral hypercalcaemia of malignancy (Suva *et al.* 1987). Since then, attention has focused on its role as an autocrine/paracrine factor that plays a role in several developmental and physiological processes. In this study we provide evidence that PTHrP may also function as a paracrine factor during endochondral ossification in the deer antler.

The findings of this study suggest that, although PTHrP is expressed in different tissues in the growing tip of the antler, its main site of synthesis in developing cartilage is likely to be the perichondrium and its target cell the chondrocyte. Several lines of

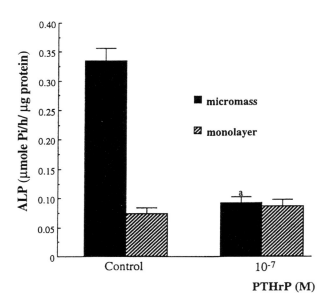

Fig. 4 Effect of PTHrP on ALP activity in cartilage cells. Cells were cultured at first passage as a monolayer or as a micromass and treated with PTHrP (10^{-7} M) for 5 days. Phosphate production is expressed as μmol/h per μg protein (mean±S.D.).
a: $P<0.001$ relative to the control.

evidence support this. First, PTHrP mRNA was expressed in perichondrium tissue, and cultured cells from this region synthesised significantly higher levels of PTHrP than those derived from cartilage. However, we observed no effect of PTHrP on the proliferation or ALP activity in cells from the perichondrium. In contrast, PTHrP stimulated proliferation and inhibited ALP activity in cultures of cells derived from cartilage. PTH/PTHrP receptor mRNA was expressed in cartilage tissue where it may mediate the effects of PTHrP on chondrocyte differentiation. These data are consistent with what is known about the role of PTHrP in controlling growth plate differentiation in the developing limb. In the rat embryo, PTHrP mRNA is expressed by perichondrial cells (Lee *et al.* 1995). Studies with transgenic mice have shown that mice homozygous for the targeted deletion of PTHrP have abnormalities in endochondral bone development; chondrocytes hypertrophy prematurely and chondrocyte division is inhibited (Karapalis *et al.* 1994). Targeted overexpression of PTHrP in the mouse growth plate results in the opposite phenotype (Weir *et al.* 1996). Immunolocalisation and *in situ* hybridisation studies are now in progress to identify which cell types in antler express PTHrP and its receptor at different stages of development.

PTHrP synthesis by the perichondrium has been shown to be positively regulated by Indian hedgehog synthesised by prehypertrophic chondrocytes (Vortkamp *et al.*

1996); it remains to be determined whether a similar feedback loop exists in the antler. In addition to the local regulation of PTHrP, it is likely that its synthesis and action is regulated by systemic factors, although very little is known about such potential mechanisms. It has recently been reported that oestrogen regulates PTHrP effects on bone cells (Kanatani *et al.* 1998), and as antler growth is closely regulated by circulating sex hormones, it provides an appropriate system for investigating how sex hormones regulate the expression and function of PTHrP.

PTHrP has been shown to be expressed in adult skin (Hayman *et al.* 1989) and may regulate hair follicle differentiation (Wysolmerski *et al.* 1994). In this study PTHrP mRNA was found to be expressed in antler velvet and was synthesised by primary cultures of velvet-derived cells. These data are consistent with the observation that PTHrP is synthesised by human keratinocytes (Merendino *et al.* 1986). Preliminary results from immunocytochemical staining of antler tissue sections confirm that PTHrP is expressed in epidermis and is associated with hair follicles. However, the absence of PTH/PTHrP receptor mRNA expression in antler velvet was unexpected as it is known to be expressed in skin during development (Lee *et al.* 1995). These findings suggest that during skin regeneration in an adult animal PTHrP may play a functional role, and this proposal requires further investigation.

In conclusion, these studies have shown that the role that PTHrP plays in regulating cell proliferation/differentiation during fetal development may be recapitulated during the regeneration of mammalian tissues.

Acknowledgements

This work was supported by the BBSRC. J S P is a Wellcome Trust Research Fellow. We thank Brendon Jackson for his expert technical assistance.

References

Goss RJ (1983) *Deer Antlers. Regeneration, Function and Evolution*. London: Academic Press.

Hayman JA, Danks JA, Ebeling PR, Moseley JM, Kemp BE & Martin TJ (1989) Expression of parathyroid hormone related protein in normal skin and in tumours of skin and skin appendages. *Journal of Pathology* **158** 293-296.

Kanatani M, Sugimoto T, Takahashi Y, Kaji H, Kitazawa R & Chihara K (1998) Estrogen via the estrogen receptor blocks cAMP-mediated parathyroid hormone(PTH)-stimulated osteoclast formation. *Journal of Bone and Mineral Research* **13** 854-862.

Karaplis AC, Luz A, Glowacki J, Bronson RT, Tybulewicz VLJ, Kronenberg HM *et al.* (1994) Lethal skelatal dysplasia from targeted disruption of the parathyroid hormone-related peptide gene. *Genes and Development* **8** 277-289.

Lee K, Deeds JD & Segre GV (1995) Expression of parathyroid hormone-related peptide and its receptor messenger ribonucleic acids during fetal development of rats. *Endocrinology* **136** 453-463.

Merendino TJ, Insogna KL, Milstone LM, Broadus AE & Stewart AF (1986) Parathyroid hormone-like protein from cultured human keratinocytes. *Science* **231** 388-390.

Price JS, Oyajobi BO, Oreffo ROC & Russell RGG (1994) Cells cultured from the growing tip of red deer antler express alkaline phosphatase and proliferate in response to insulin-like growth factor-I. *Journal of Endocrinology* **143** R9-R16.

Price JS, Oyajobi BO, Nalin AM, Frazer A, Russell RG & Sandell LJ (1996) Chondrogenesis in the regeneration antler tip in red deer: expression of collagen types I, IIA, IIB, and X demonstrated by *in situ* nucleic acid hybridization and immunocytochemistry. *Developmental Dynamics* **205** 332-347.

Schipani E, Kruse K & Juppner H (1995) A constitutively active mutant PTH-PTHrP receptor in Jansen-type metaphyseal chondrodysplasia. *Science* **268** 98-100.

Suva LJ, Winslow GA, Wettenhall RE, Hammonds RG, Moseley JM, Diefenbach-Jagger H *et al.* (1987) A parathyroid hormone-related protein implicated in malignant hypercalcemia: cloning and expression. *Science* **237** 893-896.

Vortkamp A, Lee K, Lanske B, Segre GV, Kronenberg HM & Tabin CJ (1996) Regulation of rate of cartilage differentiation by Indian hedgehog and PTH-related protein. *Science* **273** 613-622.

Weir EC, Philbrick WM, Amling M, Neff LA, Baron R & Broadus AE (1996) Targeted overexpression of parathyroid hormone-related peptide in chondrocytes causes chondrodysplasia and delayed endochondral bone formation. *Developmental Biology* **93** 1024.

Wysolmerski JJ, Broadus AE, Zhou J, Fuchs E, Milstone LM & Philbrick WM (1994) Overexpression of parathyroid hormone-related protein in the skin of transgenic mice interferes with hair follicle development. *Proceedings of the National Academy Sciences of the USA* **91** 1133-1137.

Pathophysiological effects of low dietary phosphate in pigs of southern Romania

J-L Riond[1], M Wanner[1], H Coste[2] and G Pârvu[2]

[1]Institute of Animal Nutrition, University of Zurich, Switzerland and [2]Central Laboratory for Veterinary Diagnosis, National Agency for Veterinary Health, Bucharest, Romania

Introduction

The soil of most of Romania and consequently the plants used for animal feed are known to be poor in phosphate (Pârvu 1992). The situation is exacerbated by the high phytate and low phytase content of the feed (Harland & Morris 1995, Pallauf & Rimbach 1997). In these regions, the incidence of osteomalacia and rickets is particularly high in the swine population because pigs are not able to produce appreciable amounts of phytase in the intestine, if at all (Pointillart *et al.* 1984). The homeostasis of inorganic phosphate (P_i) has been extensively studied in animals (Breves & Schröder 1991, Care 1994, Barlet *et al.* 1995) and man (Portale *et al.* 1996), but only a few *in vivo* studies have been conducted in pigs (Schröder *et al.* 1996). The present investigation was undertaken to examine the effects of low dietary phosphate on the serum concentrations of parathyroid hormone (PTH), 1,25-dihydroxyvitamin D (1,25(OH)$_2$D) and osteocalcin and on P_i homeostasis in pigs.

Materials and methods

Blood samples were examined from 82 healthy animals with known P_i status from three herds involved in a metabolic surveillance program. The calcium (Ca) and P_i contents of eight samples of the cereal-based diet fed to the different herds ranged from 1.3 to 3.9 g/kg and 1.8 to 4.0 g/kg respectively. Determination of inositol hexaphosphate and its metabolites in the food was performed by HPLC. The content of inositol hexaphosphate ranged from 1.41 to 2.99 mg/g and that of inositol penta- and di-phosphate from 0.19 to 0.62 and 0.00 to 0.06 mg/g respectively. Phytase activity ranged from 0.100 to 0.633 unit/g. $CaCO_3$ was added to the diet to meet the requirements. Serum P_i concentrations determined by the molybdate method ranged from 0.59 to 1.63 mmol/l (below normal range; $n=39$) in herd 1, 1.51 to 2.39 mmol/l (lowest part of the normal range; $n=30$) in herd 2 and 2.59 to 2.95 mmol/l (highest part of the normal range; $n=13$) in herd 3. The serum was analysed for Ca (methylthymol blue method), PTH (immunoradiometric assay; Nichols Institute Diagnostics, San Juan Capistrano, CA, USA), 1,25(OH)$_2$D (radioimmunoassay; Nichols) and osteocalcin (immunoradiometric assay; Nichols). P_i concentrations were also determined in urine and milk ($n=20$). Spearman correlation coefficients were calculated by use of the procedure CORR of the version 9.11 of the main frame-implemented SAS program.

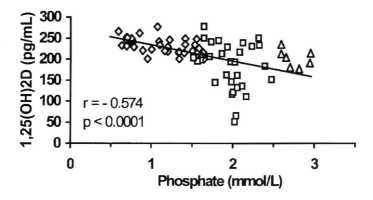

Fig. 1 Scatter plot and line of best fit of the serum concentrations of 1,25(OH)$_2$D and P$_i$ in 76 Romanian Landrace pigs from three herds with serum P$_i$ concentrations below the normal range (◊), in the lowest part of the normal range (□) and in the highest part of the normal range (Δ).

Results

Serum total Ca concentrations were within the normal range and showed no correlation with serum P$_i$. In many pigs, serum alkaline phosphatase activity was elevated. Serum concentrations of 1,25(OH)$_2$D ranged from 51.13 to 277.53 pg/ml (mean±S.D. 210.12±44.65) and may be considered high, whereas PTH concentrations ranged from 2.58 to 11.07 pg/ml (4.82±1.55), values that are low compared with reference values for pigs and other species. P$_i$ concentrations correlated negatively with 1,25(OH)$_2$D concentrations ($r=-0.574$; $P<0.0001$; Fig. 1) and positively with PTH concentrations ($r=0.617$; $P<0.0001$; Fig. 2). PTH and 1,25(OH)$_2$D concentrations correlated negatively ($r=-0.462$; $P=0.0004$). In lactating animals, no linear relationships between P$_i$ and 1,25(OH)$_2$D and between P$_i$ and PTH were observed except for that between 1,25(OH)$_2$D and PTH. Serum osteocalcin ranged from 0.08 to 3.27 ng/ml (mean±S.D. 1.16±0.58). Serum concentrations of 1,25(OH)$_2$D and osteocalcin correlated positively for the data points of herd 1 and 2 ($r=0.786$; $P<0.0001$; Fig. 3), but not for those of herd 3. Milk P$_i$ concentrations ranged from 3.10 to 7.49 mmol/l and correlated positively ($r=0.453$; $P=0.04$) with urinary P$_i$ concentrations ranging from 0.26 to 11.37 mmol/l.

Calcium Metabolism: Comparative Endocrinology

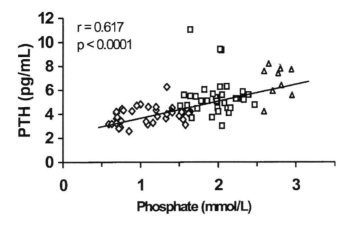

Fig. 2 Scatter plot and line of best fit of the serum concentrations of PTH and P_i in 82 Romanian Landrace pigs from three herds with serum P_i concentrations below the normal range (◊), in the lowest part of the normal range (□) and in the highest part of the normal range (Δ).

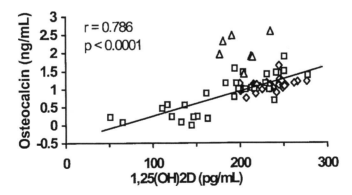

Fig. 3 Scatter plot and line of best fit of the serum concentrations of osteocalcin and $1,25(OH)_2D$ in 76 Romanian Landrace pigs from three herds with serum P_i concentrations below the normal range (◊), in the lowest part of the normal range (□) and in the highest part of the normal range (Δ).

Discussion

The elevated serum concentrations of $1,25(OH)_2D$ are caused by the low P_i content of the feed because low serum P_i induces the synthesis of $1,25(OH)_2D$ in the proximal tubules of the kidneys whereas high serum P_i decreases it (Hernandez et al. 1996, Tallon et al. 1996, Tenenhouse 1997). Intestinal absorption of P_i is in turn augmented by the increased $1,25(OH)_2D$ (Cross et al. 1990, Schröder et al. 1996). These processes are reflected by the negative correlation between serum concentrations of P_i and $1,25(OH)_2D$. Furthermore, serum P_i regulates PTH secretion by the parathyroid gland (Almaden et al. 1996, Wang et al. 1996). High serum P_i increases PTH secretion whereas low serum P_i decreases it. These processes are reflected by the positive correlation between P_i and PTH serum concentrations. The relatively low serum PTH concentrations are probably due to the high serum $1,25(OH)_2D$ levels, as $1,25(OH)_2D$ is known to inhibit PTH secretion (Funahashi et al. 1998). The increased $1,25(OH)_2D$ and decreased PTH are thus the response of the interacting regulatory loop for P_i homeostasis during dietary P_i deficiency. During lactation, other hormones such as PTH-related peptide and prolactin or a change in the affinity of the vitamin D receptor may be involved because linear relationships between P_i and $1,25(OH)_2D$ and between P_i and PTH were not observed. As P_i deficiency induces hypercalcemia (Schröder et al. 1996, Hoshino et al. 1998), the apparent lack of a negative correlation between serum total Ca and P_i concentrations was unexpected. The clear relationship between osteocalcin and $1,25(OH)_2D$ may be explained by stimulation of osteocalcin biosynthesis in osteoblasts by $1,25(OH)_2D$ which has previously been demonstrated *in vitro* and *in vivo* (Price & Baukol 1980, 1981). The positive correlation between milk and urine P_i suggests that the mechanisms of hormonal regulation of P_i reabsorption in the proximal tubule may be, at least in part, similar to those of P_i secretion into milk.

References

Almaden Y, Canalejo A, Hernandez A, Ballesteros E, Garcia-Navarro S, Torres A & Rodriguez M (1996) Direct effect of phosphorus on PTH secretion from whole rat parathyroid glands *in vitro*. *Journal of Bone and Mineral Research* **11** 970-976.

Barlet JP, Davicco MJ & Coxam V (1995) Physiologie de l'absorption intestinale du phosphore chez l'animal. *Reproduction Nutrition Development* **35** 475-489.

Breves G & Schröder B (1991) Comparative aspects of gastrointestinal phosphorus metabolism. *Nutrition Research Reviews* **4** 125-140.

Care AD (1994) The absorption of phosphate from the digestive tract of ruminant animals. *British Veterinary Journal* **150** 197-205.

Cross HS, Debiec H & Peterlik M (1990) Mechanism and regulation of intestinal phosphate absorption. *Mineral and Electrolyte Metabolism* **16** 115-124.

Funahashi H, Tanaka Y, Imai T, Wada M, Tsukamura K, Hayakawa Y et al. (1998) Parathyroid hormone suppression by 22-oxacalcitriol in the severe parathyroid hyperplasia. *Journal of Endocrinological Investigation* **21** 43-47.

Harland BF & Morris ER (1995) Phytate: a good or a bad food component? *Nutrition Research* **15** 733-754.

Hernandez A, Conception MT, Rodriguez M, Salido E & Torres T (1996) High phosphorus diet increases preproPTH mRNA independent of calcium and calcitriol in normal rats. *Kidney International* **50** 1872-1878.

Hoshino H, Kushida K, Takahashi M, Koyama S, Yamauchi H & Inoue T (1998) Effects of low phosphate intake on bone and mineral metabolism in rats: evaluation by biochemical markers and pyridium cross-link formation in bone. *Annals of Nutrition and Metabolism* **42** 110-118.

Pallauf J & Rimbach G (1997) Nutritional significance of phytic acid and phytase. *Archiv für Tierernährung (Archives of Animal Nutrition)* **50** 301-319.

Pârvu G (1992) *Supravegherea Nutritional Metabolica A Animalelor.* Bucharest, Romania: Ceres.

Pointillart N, Fontaine N & Thomasset M (1984) Phytate phosphorus utilization and intestinal phosphatases in pigs fed low phosphorus: wheat or corn diets. *Nutrition Report International* **29** 473-483.

Portale AA, Halloran BP, Morris RC & Lonergan ET (1996) Effect of aging on the metabolism of phosphorus and 1,25-dihydroxyvitamin D in healthy men. *American Journal of Physiology* **270** E483-E490.

Price PA & Baukol SA (1980) 1,25-Dihydroxyvitamin D increases synthesis of the vitamin D-dependent bone protein by osteosarcoma cells. *Journal of Biological Chemistry* **225** 11660-11663.

Price PA & Baukol SA (1981) 1,25-Dihydroxyvitamin D_3 increases serum levels of the vitamin K-dependent bone protein. *Biochemical and Biophysical Research Communications* **99** 928-935.

Schröder B, Breves G & Rodehutscord M (1996) Mechanisms of intestinal phosphorus absorption and availability of dietary phosphorus in pigs. *Deutsche Tierärtzliche Wochenschrift* **103** 197-236.

Tallon S, Berdud I, Hernandez A, Conception MT, Almaden Y, Torres A, *et al.* (1996) Relative effects of PTH and dietary phosphorus on calcitriol production in normal and azotemic rats. *Kidney International* **49** 1441-1446.

Tenenhouse HS (1997) Cellular and molecular mechanisms of renal phosphate transport. *Journal of Bone and Mineral Research* **12** 159-164.

Wang Q, Paloyan E & Parfitt AM (1996) Phosphate administration increases both size and number of parathyroid cells in adult rats. *Calcified Tissue International* **58** 40-44.

Acceleration of rat femoral fracture healing by a synthetic thrombin peptide

D J Simmons[1], J Yang[1], S Yang[1], L X Bi[1], W L Buford[1], R T Turner[2], R Crowther[3,4] and D H Carney[3,4]

[1]Department of Orthopaedic Surgery and Rehabilitation, University of Texas Medical Branch, Galveston, Texas 77555-0892, USA, [2]Orthopaedic Research, Mayo Clinic and Foundation, Rochester, Minnesota 55905, USA, [3]Department of Human Biological Chemistry and Genetics, University of Texas Medical Branch, Galveston, Texas 77555-0892, USA and [4]Chrysalis Biotechnology, Galveston, Texas 77550, USA

Introduction

The current therapeutic options for malunions are limited to bioelectrical (Connoly 1979, Hinsenkamp *et al.* 1987) and biomechanical approaches (Burny & Donkerwolcke 1987), which have uncertain outcomes. Growth factor therapy is now possible; basic fibroblast growth factor (bFGF; Kawaguchi *et al.* 1994), endothelial cell stimulating angiogenic factor (ESAF; Kurdy *et al.* 1996), transforming growth factor-β (Lind *et al.* 1993) and nerve growth factor (Grills *et al.* 1997) are available for research. Osteogenic proteins (OP-1 and OP-2), members of the bone morphogenetic protein family, are currently undergoing clinical trials. However, despite their promise, the use of recombinant growth factors involves a high cost, a potential for overstimulation of specific target cells leading to partial healing or possible overgrowth, and a potential for systemic effects or latent transformation of cells.

Here, we explore the use of a 23-amino acid fragment of the human thrombin molecule (TP508), which binds to the non-proteolytically activated receptor for thrombin (Glenn *et al.* 1988). TP508 is best known to accelerate epidermal repair (Carney *et al.* 1992) through its angiogenic and mitogenic activities (Carney *et al.* 1992, Kim *et al.* 1994). In fact, bone repair involves some of the same basic cell types and processes. Specifically, clot formation and platelet activation are initiated by thrombin. Dead tissue is removed by neutrophils and macrophages that invade the clot and surrounding tissue. Capillary revascularization permits entry and modulation of fibroblast-like cells to chondrogenic and osteogenic elements. The degree of vascularity at the repair site is central to these processes (Trueta 1974). It is likely, then, that a factor such as TP508 could be efficacious in accelerating fracture repair. In the present work we report the investigation of this hypothesis in a rat femoral closed transverse fracture model.

Methods

Fracture model

Effect of age

Under general anesthesia, the left femurs of young (2 months old; $n=160$) and skeletally mature (>8 months old; $n=160$) male Sprague-Dawley rats were roded with B&S gauge 22 surgical steel (Ethicon) inserted proximal to the intercondylar notch, and fractured transversely at the level of the midshaft by bending to failure (Bonnarens & Einhorn 1984). In randomly assigned treatment groups ($n=10$), four groups received a single postoperative injection of 1.0 µg TP508 dissolved in 0.1 ml sterile PBS directly into the fracture site. Fractures in two additional groups were injected with bFGF (at 1.0 or 10.0 ng) on day 3. Control group fractures were injected with 0.1 ml PBS.

Ten animals in each of the treatment groups and their controls were killed (pentobarbital 100 mg/kg) 4 weeks after the fracture. The fractured and intact contralateral femurs were recovered at autopsy, photographed and radiographed to visualize the position of the fracture. Bones with unequivocal diaphyseal fractures were wetted with sterile saline and stored at −90°C for subsequent biomechanical analysis.

A second experiment also assessed the effect of TP508 treatment on fracture healing over an 8 week period in four groups of 15 control and 15 experimental rats that were 2 months old at the time of fracture. Here, the femurs were roded with stiffer 0.35 inch Steinman pins (Howmedica, Rutherford, NJ, USA); the protocols were otherwise identical. At autopsy, the femurs from three animals in each group were processed for histological examination and 12 were frozen for biomechanical analysis. All experimental procedures were approved by the UTMB Institutional Animal Use and Care Committee (ACUC Protocol no. 95-12-077).

Biomechanics

Femurs were tested in a three-point bending apparatus (Bak & Jensen 1992), using an MTS-858 Minibionix machine (MTS, Minneapolis, MN, USA). The ram was advanced at 0.1 mm/s until the absolute breaking strength had been reached. Displacement data (mm) were recorded at 1 s intervals. Tissue strength at the ultimate failure point was expressed in Newtons (N).

Histology

Bone was fixed in 10% neutral formalin, embedded undemineralized in methylmethacrylate, and sectioned parasagittally at 2-3 µm. The sections were stained with toluidine blue. At the light microscopic level (×63), the toluidine blue-stained fracture tissue was displayed on a TV monitor, and the percentage distributions of cartilage, fibrocartilage, trabecular bone and cortical bone in the callus were calculated using a computerized software package (Optimas Corp., Bothell, WA, USA).

Statistical analysis

The data were expressed as the mean±S.E.M. The differences between the means were analyzed by a non-parametric two-tailed t test, F test and Wilcoxon rank-order statistics. Differences between means at the 5% confidence level ($P<0.05$) were considered to be statistically significant.

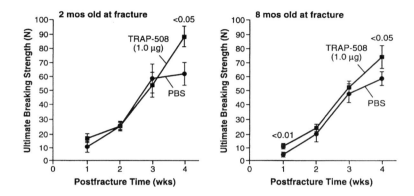

Fig. 1 Time course of repair of a closed femoral fracture in skeletally immature (2 months (mos) old at fracture) and mature (8 months old at fracture) rats after an acute injection of 1.0 µg TP508 (TRAP-508; ■) or PBS (●; controls) into their fresh bone lesions. The changes in bone breaking strength (N) showed that TP508 enhanced fracture healing throughout the 4 week study.

Results

Effects of TP508 and bFGF on fracture healing in skeletally immature and mature rats

Aging study

Measurements made at 4 weeks after the fracture showed that intra- and peri-osseous TP508 and bFGF injection were effective in producing a statistically significant increase in breaking strength. TP508-treated fractures from young and old rats were respectively 43% ($P<0.05$) and 26% ($P<0.05$) stronger than their vehicle-treated controls (Fig. 1). In contrast, 10 ng bFGF significantly increased breaking strength in mature rats ($P<0.05$), but not in young animals (Fig. 2). No effect was observed at the 1.0 ng bFGF level in either age group.

The 8 week follow-up study confirmed the efficacy of TP508 treatment on the rate of fracture healing in terms of the breaking strength and the restoration of normal tissue quality (Fig. 3). TP508-treated fractures were as strong at 2 weeks as the untreated PBS fractures at 4 wks, showing an early twofold enhancement during the first month ($P<0.001$). The benefit of TP508 was also clear in terms of shortening the time required to restore normal bone breaking strength (TP508, 6-7 weeks; controls, >8 weeks).

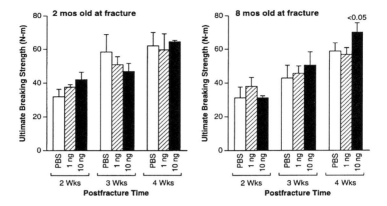

Fig. 2 Time course of repair of a closed femoral fracture in skeletally immature (2 months (mos) old at fracture) and mature (8 months old at fracture) rats after an acute injection of 1.0 (hatched bar) or 10 (solid bar) ng bFGF or PBS (open bar; controls) into their fresh bone lesions. The changes in bone breaking strength (N) showed that bFGF enhanced healing only in 8-month-old rats throughout the 4 week study.

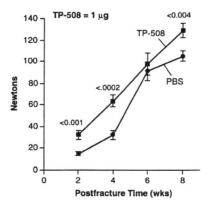

Fig. 3 Time course of repair of a closed femoral fracture in young rats after an acute injection of 1.0 μg TP508 (■) or PBS (●; controls) into their fresh bone lesions. The breaking strength of fractures treated acutely with TP508 was more nearly normal after 6-8 weeks than their PBS-treated controls.

Calcium Metabolism: Comparative Endocrinology

Histomorphometry

In terms of relative tissue volumes, there were few histotypic differences between TP508-treated animals and their controls. However, there was significant retention of hyaline cartilage in the PBS-treated controls in the first month after fracture, indicating that TP508 accelerated the time to ossification by some 2 weeks (Fig. 4).

Discussion

These results show that TP508 is a potent stimulator of fracture healing in a rat model, as demonstrated by enhanced mechanical strength and accelerated progression of the healing process documented by histomorphology. A single dose of 1 µg TP508

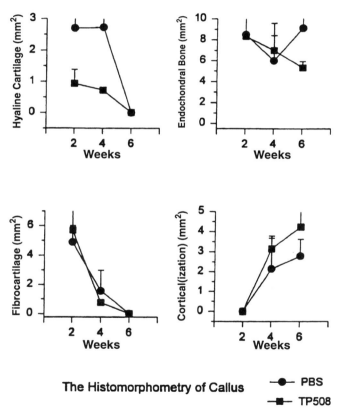

Fig. 4 Plots showing the time-course changes in the distribution of tissue types in the healing fractures of skeletally immature male rats after an acute injection of 1.0 µg TP508 (■) or PBS (●; controls) into their fresh bone lesions. The more rapid loss of cartilage in the TP508-treated bones suggests a local increase in skeletal maturation.

doubled the initial rate at which mechanical strength was returned to the limb. In addition, TP508 proved more effective than bFGF, the current gold standard in laboratory studies of fracture repair.

The mechanisms by which TP508 might stimulate wound healing involve acceleration of osteoblast and marrow stromal cell proliferation, but such effects were relatively modest (30-40%) and occurred only in the presence of high unphysiological concentrations (10^{-4} M; data not shown). As mentioned in the Introduction, it seems that, although the pharmacokinetics of TP508 at the fracture site are unknown, the consequences favorable for fracture healing are likely to include neutrophil chemotaxis, fibroblast proliferation (Kim et al. 1994) and angiogenesis (Carney et al. 1992). The latter is the critical step in the progress to bone union and return of mechanical strength (Simmons 1980). A number of small angiogenic factors have been identified (Oderdra & Weiss 1991), and, of these, ESAF has been shown to increase in concentration in patients with fractures (Kurdy et al. 1996). The rise in ESAF levels was biphasic, with an initial increase at day 2 after injury, followed by a larger more sustained increase that peaked on days 8-10 after injury, which was possibly associated with reorganization of the fracture hematoma and vascular invasion. Therefore, the administration of TP508 at the time of injury, or shortly after, may significantly accelerate neovascularization and thus accelerate the whole fracture repair process.

Acknowledgements

The work was supported by a grant from the University of Texas Center on Aging.

References

Bak B & Jensen KS (1992) Standaridization of tibial fractures in the rat. *Bone* **13** 289-295.

Bonnarens F & Einhorn TA 1984 Production of a standard closed fracture in laboratory animal bone. *Journal of Orthopaedic Research* **2** 97-101.

Burny F & Donkerwolcke M (1987) Elastic fixation of fractures: biomechanics of fracture healing. In *Fracture Healing*, pp. 123-137. Ed JM Lane. New York: Churchill-Livingstone.

Carney DH, Mann R, Redin WR, Pernia SD, Berry D, Heggers JP, Hayward PG, Robson MD, Christie J, Annable C, Fenton JW & Glenn KC (1992) Thrombin and synthetic thrombin receptor-activating peptides enhance incisional wound healing and neovascularization. *Journal of Clinical Investigations* **89** 1469-1477.

Connolly JF (1979) Healing curve of fractures and the effect of bone growth stimulation. In *Electrical Properties of Bone and Cartilage. Experimental Effects and Clinical Applications*, pp 547-562. Eds CT Brighton, J Black & SR Pollack. New York: Grune & Stratton.

Glenn KC, Herbosa GJ & Carney DH (1988) Synthetic peptides bind to high-affinity thrombin receptors and modulate thrombin mitogenesis. *Peptide Research* **2** 65-93.

Grills BL, Schuijers JA & Ward AR (1997) Topical application of nerve growth factor improves fracture healing in rats. *Journal of Orthopaedic Research* **15** 235-242.

Hinsenkamp MG, Rooze M & Tuerlinckx B (1979) Biophysical and biological intervention: future directions. In *Fracture Healing*, pp. 267-274. Ed JM Lane. New York: Churchill-Livingstone.

Kawaguchi H, Kurokawa T, Hanada K, Hiyama Y, Tamura M, Ogata E & Matsumoto T (1994) Stimulation of fracture repair by recombinant basic fibroblast growth factor in normal and streptozotocin-diabetic rats. *Endocrinology* **135** 774-778.

Kim D, Wang F, Ramakrishnan S, Scott DL, Hensler TM, Thompson WC & Carney DH (1994) Fibroblasts defective in thrombin mitogenesis exhibit normal expression and activation of the proteolytically activated receptor for thrombin: requirement for a second signaling pathway. *Journal of Cell Physiology* **160** 573-584.

Kurdy NMG, Weiss JB & Bate A (1996) Endothelial stimulating angiogenic factor in early fracture healing. *Injury* **27** 143-145.

Lind M, Schumaker B, Soballe K, Keller J, Flemming M & Bunger C (1993) Transforming growth factor-β enhances fracture healing in rabbit tibiae. *Acta Orthopedica Scandinavica* **64** 553-556.

Odedra R & Weiss JB (1991) Low molecular weight angiogenesis factors. *Pharmacological Therapy* **49** 111-116.

Simmons DJ (1980) Fracture healing. In *Fundamental and Clinical Bone Physiology*, pp. 283-330. Ed MR Urist. Philadelphia: Lippincott.

Trueta J (1974) Blood supply and the rate of healing of tibial fractures. *Clinical Orthopedics and Related Research* **105** 11-15.

Regulation of fetal-placental calcium metabolism

C S Kovacs[1], B Lanske[2], N R Manley[3] and H M Kronenberg[4]

[1]Faculty of Medicine - Endocrinology, Health Sciences Centre, Memorial University of Newfoundland, St John's, Newfoundland A1B 3V6, Canada, [2]Molecular Endocrinology, Max Planck Institute for Biochemistry, 82152 Martinsried, Germany, [3]Institute for Molecular Medicine and Genetics, Medical College of Georgia, Augusta, Georgia 30912, USA and [4]Endocrine Unit, Massachusetts General Hospital and Harvard Medical School, Boston, Massachusetts 02114, USA

Introduction

Calcium and bone metabolism has been uniquely adapted to meet the specific needs during fetal development; this subject has been reviewed in depth elsewhere with detailed references (Kovacs & Kronenberg 1997*b*). We have explored the regulation of normal fetal calcium physiology through the use of gene targeting models. These techniques permit the study of fetal models that cannot be created by surgical or pharmacological techniques, such as fetal mice whose parathyroids produce parathyroid hormone (PTH) but not PTH-related protein (PTHrP), and vice versa. This paper will briefly review what we have learned from these studies, in the context of what is already known about fetal calcium homeostasis.

Fetal hypercalcemia

In mammals, the fetal blood calcium (total and ionized) is maintained at a higher level than in the maternal circulation. The rodent fetus has a remarkable ability to maintain this higher level of blood calcium despite severe maternal hypocalcemia. In adults, the calcium-sensing receptor (CaSR), which directly regulates PTH secretion, normally sets the blood calcium level. Using genetically engineered mice that lack the CaSR gene, we have recently shown that the murine fetus sets its blood calcium level independently of the ambient maternal calcium level (Kovacs *et al.* 1998*a*).

Fetal calciotropic hormones

Several features of calciotropic hormone levels in the fetus provide further evidence that calcium homeostasis is regulated differently from the adult. The fetal parathyroid glands contain abundant PTH mRNA and are capable of synthesizing PTH, but the level of PTH in the fetal circulation is generally very low near term. Our studies on fetal mice with heterozygous and homozygous disruption of the CaSR indicate that PTH may be suppressed by a normal CaSR in response to the increased fetal blood calcium (Kovacs *et al.* 1998*a*). Since the fetal blood calcium is elevated at the same time that the PTH level is suppressed, factors other than PTH must regulate the blood calcium level.

Fetal 1,25-dihydroxyvitamin D (1,25-D) levels are very low compared with the adult, and thus 1,25-D is not likely responsible for the high fetal blood calcium level. In fact, absence of the vitamin D receptor itself has little or no apparent effect on the fetal blood calcium and skeletal development (Li et al. 1997). Fetal calcitonin levels are higher than maternal levels but, as with adults, the importance of calcitonin as a calciotropic hormone is uncertain.

PTHrP has actions that may be unique to fetal calcium homeostasis. PTHrP may actually be processed into separate circulating fragments, each of which may have different functional roles and receptors. Although normally undetectable in the adult circulation, human and animal studies have found that PTHrP levels are up to 15-fold higher than the simultaneous levels of PTH at term. We confirmed that PTHrP has an important role in maintaining the normal raised fetal blood calcium by observing a lowered blood calcium (equal to the maternal blood calcium) in PTHrP-null fetuses (Kovacs et al. 1996). Serum PTH levels are sharply increased in these PTHrP-null fetuses, but the blood calcium remains low (Kovacs et al. 1997).

Placental calcium transport

The fetus and placenta must obtain sufficient calcium to mineralize the skeleton, and maintain an extracellular level of calcium that is physiologically appropriate for fetal tissues (i.e. for cell membrane stability, blood coagulation, etc). The bulk of placental calcium transfer occurs late in gestation, such that 80% occurs in the third trimester in humans, while 96% occurs in the last 5 days of gestation in the rat.

Maternal regulation of placental calcium transport

Maternal hormones might influence fetal-placental calcium transport by raising or lowering the ambient maternal calcium level, and by direct effects on the placenta. However, the published literature indicates that a normal rate of maternal-to-fetal calcium transfer can usually be maintained despite the presence of maternal hypocalcemia or hormone deficiencies.

Fetal regulation of placental calcium transport

PTH, 1,25-D, calcitonin, and PTHrP have been studied to varying degrees regarding their potential roles in regulating placental calcium transfer. The parathyroid glands have been extensively studied, although it is only most recently that it has been recognized that the actions of the parathyroids in fetal life may not be through PTH alone. Fetal thyroparathyroidectomy in sheep resulted in a lower fetal blood calcium, and reduced transport of calcium across the artificially perfused placentas (Care et al. 1986). These findings confirmed that the fetal parathyroids have a critical role in maintaining the fetal blood calcium and placental calcium transfer. Blood from intact fetuses restored calcium transport, as did PTHrP, while PTH itself was without effect (Rodda et al. 1988). This was the first evidence that the fetal parathyroids might regulate placental calcium transfer through a factor other than PTH.

More recently, the studies that we have completed in intact fetal mice support the hypothesis that fetal PTHrP stimulates placental calcium transport *in vivo*. In addition to lowered blood calcium, PTHrP-null fetal mice have reduced placental calcium transfer (Kovacs et al. 1996). Treatment of these PTHrP-null fetuses in utero with certain fragments of PTHrP (PTHrP 1-86 or PTHrP 67-86) acutely increased the rate of placental calcium transfer, while PTHrP 1-34 and intact PTH had no effect. This suggests that fetal PTHrP regulates placental calcium transfer through a receptor (as yet uncloned) which is distinct from the known PTH/PTHrP receptor. The reason for this conclusion is twofold: first, the PTH/PTHrP receptor is activated equally by amino-terminal PTH and PTHrP, but both peptides failed to stimulate placental calcium transfer in the PTHrP-null fetus; and second, PTHrP 67-86 (the 'mid-molecule' of PTHrP) stimulated placental calcium transfer, but this fragment does not bind to or activate the known PTH/PTHrP receptor. This hypothesis was further supported by our studies of PTH/PTHrP receptor-null fetuses. These fetuses are also hypocalcemic, but placental calcium transfer is increased, perhaps due to up-regulation of the PTHrP mid-molecule and its effect (Kovacs et al. 1996).

Although the studies in fetal sheep suggested that parathyroid-derived PTHrP regulates placental calcium transfer, no confirmatory measurements of PTHrP levels were made. Our studies in the PTHrP-null fetus do not address the relative importance of parathyroid-derived PTHrP, since PTHrP was absent from all tissues of the PTHrP-null fetus. It may well be that PTHrP derived from the placenta regulates placental calcium transfer.

We are currently utilizing the Hoxa3 gene knockout model (Manley & Capecchi 1995) to assess the relative importance of PTHrP derived from the parathyroids versus PTHrP derived from the placenta in regulating these processes. We have recently shown that fetal mice lacking parathyroid glands (Hoxa3-null fetuses) have reduced placental ^{45}Ca transfer, similar to the value observed in PTHrP-null fetuses (Kovacs et al. 1998b). Since Hoxa3-null fetuses and PTHrP-null fetuses have a similar reduction in placental calcium transfer, it is possible that Hoxa3-null fetal mice also lack circulating PTHrP, but this remains to be determined.

Molecular mechanisms of placental calcium transport

The site of active calcium transport is likely in the trophoblast cells and in the endoderm of the intraplacental yolk sac in rodents. These cells are in closest proximity to the maternal circulation, and the intraplacental yolk sac cells express particularly high levels (mRNA and protein) of a Ca^{2+}-ATPase (the putative 'calcium pump') and high levels of calbindin$_{9K}$-D (a calcium-binding protein). Little work has been done to establish the role of the Ca^{2+}-ATPase and calbindin$_{9K}$-D in placental calcium transfer, or to verify that the cells that express these molecules are, indeed, the cells that transport calcium. We have examined PTHrP-null placentas and have found reduced calbindin$_{9K}$-D mRNA and protein levels in the intraplacental yolk sac (Kovacs & Kronenberg 1997a). The relevance of these findings to the reduced placental calcium transfer seen in PTHrP-null fetuses remains to be determined.

Fetal parathyroids

The fetal parathyroids appear to contribute to calcium homeostasis by secretion of both PTH and PTHrP. Although the postulated importance of PTHrP has been cited, other evidence suggests that PTH may also be important for normal fetal calcium homeostasis. PTH can be secreted by the fetal parathyroids at higher rates in response to acute hypocalcemia. Further, fetal mice lacking the PTH/PTHrP receptor gene (and, therefore, the effects of both PTH and PTHrP on this receptor) have significantly lower blood calcium than do fetal mice that lack PTHrP alone (Kovacs et al. 1996). Also, Hoxa3-null mice that lack parathyroid glands (and, presumably, parathyroid-derived PTH and PTHrP) have a blood calcium level lower than that of PTHrP-null fetuses, and equal to that of PTH/PTHrP receptor-null fetuses (Kovacs et al. 1998b).

The evidence that the fetal parathyroids even make PTHrP is not conclusive. Studies in the parathyroids of fetal sheep and calves have noted immunoreactivity (by immunohistochemistry and Western blot) to both mid-molecular and amino-terminal fragments of PTHrP in the parathyroids (Abbas et al. 1990, MacIsaac et al. 1991). Taken together with the finding of reduced PTHrP-responsive placental calcium transport in parathyroidectomized fetal sheep, it has been postulated that the parathyroids regulate placental calcium transport through the production of PTHrP. However, a recent study of the parathyroids of fetal rats found abundant production of PTH, but no detectable evidence of PTHrP mRNA or the protein (Tucci et al. 1996). This latter study underscores the need for further clarification of the postulated role of parathyroid-produced PTHrP in regulating placental calcium transfer and the fetal blood calcium.

Fetal kidneys and amniotic fluid

The fetal kidneys may partly regulate fetal calcium homeostasis, by adjusting the relative reabsorption and excretion of calcium and phosphate by the renal tubules in response to the filtered load and other factors, such as PTHrP and/or PTH. The fetal kidneys may also participate by synthesizing 1,25-D. Since calcium excreted by the kidneys can be recycled to the fetus by swallowing the amniotic fluid, the relative importance of renal calcium handling in regulating the fetal blood calcium is uncertain and not easily studied.

Fetal skeleton

The recent observations that we have made suggest a role for the skeleton in fetal calcium homeostasis. The ionized calcium of PTH/PTHrP receptor-null fetal mice is lower than that of PTHrP-null fetal mice, despite the fact that placental calcium transport is supranormal in PTH/PTHrP receptor-null fetuses, and subnormal in PTHrP-null fetuses (Kovacs et al. 1996). Lack of bone responsiveness to amino-terminal PTH and PTHrP may well, therefore, contribute to the hypocalcemia in mice without PTH/PTHrP receptors. We have also found that disruption of the CaSR in fetal mice raises the ionized calcium higher than normal, and this is maintained apparently through increased PTH-stimulated bone resorption or decreased incorporation of calcium into bone (Kovacs et al. 1998a).

Intact fetal parathyroid glands may be needed for normal skeletal development, since thyroparathyroidectomy in fetal lambs causes decreased skeletal calcium content and rachitic changes (Aaron *et al.* 1992). These effects could be partly reversed or prevented by fetal calcium and phosphate infusions; thus, much of the effect of fetal parathyroidectomy was caused by a decrease in blood levels of calcium and phosphate. Therefore, functioning fetal parathyroids (and, therefore, parathyroid gland-produced PTH and/or PTHrP) are required for normal fetal bone resorption and mineralization.

Conclusions

We propose the following integrated model of normal fetal calcium homeostasis. The fetal blood calcium is set at a level higher than the maternal level through the actions of PTHrP (among other potential factors and other actions of PTHrP). The parathyroid CaSR responds appropriately to this increased level of calcium, and suppresses the synthesis and release of PTH from the parathyroids. 1,25-D synthesis and secretion are, in turn, suppressed due to the effects of low PTH, and high blood calcium and phosphate. The parathyroids may play a central role by producing PTHrP, or may be responding to the effects of PTHrP produced by the placenta. The fetal blood calcium is maintained not only by flux of calcium across the placenta from the mother, but by contributions from fetal skeleton and kidney. The incorporation of calcium into the growing fetal skeleton may well be regulated by other factors in addition to PTH and PTHrP.

References

Aaron JE, Abbas SK, Colwell A, Eastell R, Oakley BA, Russell RG *et al.* (1992) Parathyroid gland hormones in the skeletal development of the ovine foetus: the effect of parathyroidectomy with calcium and phosphate infusion. *Bone and Mineral* **16** 121-129.

Abbas SK, Pickard DW, Illingworth D, Storer J, Purdie DW, Moniz C *et al.* (1990) Measurement of parathyroid hormone-related protein in extracts of fetal parathyroid glands and placental membranes. *Journal of Endocrinology* **124** 319-325.

Care AD, Caple IW, Abbas SK & Pickard DW (1986) The effect of fetal thyroparathyroidectomy on the transport of calcium across the ovine placenta to the fetus. *Placenta* **7** 417-424.

Kovacs CS & Kronenberg HM (1997*a*) PTHrP regulates placental calbindin-D-9K expression: evidence from the PTHrP gene 'knockout' model (Abstract). *Journal of Bone and Mineral Research* **12** (Suppl 1) S212.

Kovacs CS & Kronenberg HM (1997*b*) Maternal-fetal calcium and bone metabolism during pregnancy, puerperium and lactation. *Endocrine Reviews* **18** 832-872.

Kovacs CS, Lanske B, Hunzelman JL, Guo J, Karaplis AC & Kronenberg HM (1996) Parathyroid hormone-related peptide (PTHrP) regulates fetal-placental calcium transport through a receptor distinct from the PTH/PTHrP receptor. *Proceedings of the National Academy of Sciences of the USA* **93** 15233-15238.

Kovacs CS, Lanske B, Byrne M, Krane SM & Kronenberg HM (1997) Altered interstitial collagenase expression in the tibias of PTHrP gene-ablated and PTH/PTHrP receptor gene-ablated fetal mice (Abstract). *Journal of Bone and Mineral Research* 12 (Suppl 1) S116.

Kovacs CS, Ho-Pao CL, Hunzelman JL, Lanske B, Fox J, Seidman JG *et al.* (1998*a*) Regulation of murine fetal-placental calcium metabolism by the calcium-sensing receptor. *Journal of Clinical Investigation* **101** 2812-2820.

Kovacs CS, Manley NR & Kronenberg HM (1998*b*) *Hoxa3* knockout mice are hypocalcemic and have reduced placental calcium transfer (Abstract). In *Proceedings of the 80th Annual Meeting of the Endocrine Society*, New Orleans, LA, 24-27 June 1998, pp 91.

Li YC, Pirro AE, Amling M, Delling G, Baron R, Bronson R *et al.* (1997) Targeted ablation of the vitamin D receptor: an animal model of vitamin D dependent rickets type II with alopecia. *Proceedings of the National Academy of Sciences of the USA* **94** 9831-9835.

MacIsaac RJ, Caple IW, Danks JA, Diefenbach-Jagger H, Grill V, Moseley JM *et al.* (1991) Ontogeny of parathyroid hormone-related protein in the ovine parathyroid gland. *Endocrinology* **129** 757-764.

Manley NR & Capecchi MR (1995) The role of *Hoxa-3* in mouse thymus and thyroid development. *Development* **121** 1989-2003.

Rodda CP, Kubota M, Heath JA, Ebeling PR, Moseley JM, Care AD *et al.* (1988) Evidence for a novel parathyroid hormone-related protein in fetal lamb parathyroid glands and sheep placenta: comparisons with a similar protein implicated in humoral hypercalcaemia of malignancy. *Journal of Endocrinology* **117** 261-271.

Tucci J, Russell A, Senior PV, Fernley R, Ferraro T & Beck F (1996) The expression of parathyroid hormone and parathyroid hormone-related protein in developing rat parathyroid glands. *Journal of Molecular Endocrinology* **17** 149-157.

Endocrinology of calcium metabolism in rabbits

R Brommage

Section on Comparative Medicine, Department of Pathology, Wake Forest University School of Medicine, Medical Center Boulevard, Winston-Salem, North Carolina 27157-1040, USA

Introduction

The goal of this brief review is to describe the known information on Ca metabolism and its regulation in rabbits, with emphasis on comparisons with other mammals. Given space limitations, the numerous relevant citations will not be provided. However, a WORD document listing over 250 references will be sent to anyone requesting this information by E-mail (brommage@wfubmc.edu).

Dietary Ca requirement and absorption

To achieve normal bone mass, growing rabbits require a diet containing at least 0.2% Ca as diets with lower Ca contents lead to reduced bone mineral density (Gilsanz *et al* 1991) and cortical width in the spine (Wu *et al* 1990). Several studies have shown that rabbits absorb a large percentage of ingested Ca with minimal declines in fractional absorption as the dietary Ca content is increased to high levels. The absorbed Ca not utilized for skeletal mineralization is excreted both in urine and as endogenous fecal Ca.

Serum Ca homeostasis and parathyroid hormone action

The serum Ca concentration in rabbits is ~3.25 mM, a value 30% higher than that observed in other mammals. Ionized Ca levels are similarly increased, indicating that the binding of Ca to serum protein is normal (Warren *et al* 1989). The rabbit parathyroid gland Ca receptor that signals the parathyroid cell to secrete PTH has a normal binding affinity for Ca consistent with an amino acid sequence similar to that of other mammals (Butters *et al* 1997). Hypocalcemia stimulates PTH secretion and parathyroidectomy produces severe hypocalcemia (Tan *et al* 1987). The receptor for PTH has been characterized in kidney cells (Kremer *et al* 1982). As described in more detail below, PTH promotes renal tubular Ca reabsorption. Thus, although rabbits are normally hypercalcemic relative to other mammals, the reason for this high serum Ca level has not been identified but does not appear to involve a derangement in PTH secretion or action. Numerous factors have been shown to influence serum Ca levels in rabbits. In addition to conventional influences, severe hypocalcemia is produced by injections of protamine (which presumably mimics Ca in activating the parathyroid cell Ca receptor) and asparaginase (which causes the destruction of parathyroid cells).

Vitamin D metabolism and action

The metabolism and Ca-regulatory actions of vitamin D in rabbits are typical of mammals. Circulating levels of calcitriol are increased in response to a low Ca diet. The plasma vitamin D-binding protein has been purified and sequenced and determined to have a half-life of 1.7 days (Haddad *et al* 1981). Rabbit intestine and chondrocytes contain the vitamin D receptor. Calcitriol infusion markedly depresses serum PTH levels. Although their vitamin D dependence has not been determined, calbindin D9K and calbindin D28K are present in the intestine and kidney respectively.

Calcitonin

The rabbit thyroid parafollicular cells secrete calcitonin in response to hypercalcemia and calcitonin acts to lower serum Ca levels. Rabbit calcitonin and its receptor have both been sequenced.

Intestinal ca absorption

When measured by classical balance techniques, fractional intestinal Ca absorption in the rabbit is usually observed to be about 50%. Although active Ca transport against a concentration gradient has been demonstrated in the rabbit duodenum using the everted gut sac technique, more rigorous studies using voltage-clamped tissue in a Ussing chamber set-up do not appear to have been performed. There is no reason to suspect that active transport of Ca using a calcitriol-dependent mechanism does not occur. In addition to the calcitriol receptor and calbindin D9K, the rabbit intestine also contains the plasma membrane Ca^{2+}-ATPase believed to be responsible for transporting Ca^{2+} across the basolateral membrane of the enterocyte. This Ca^{2+} pump is found predominantly in the rabbit duodenum and proximal colon, sites having the greatest active transport capacities in other species. Nonetheless, Ca^{2+}-ATPase-dependent active Ca^{2+} transport has been demonstrated in basolateral membranes derived from the distal colon.

Renal Ca excretion

Rabbit urine is normally cloudy due to the presence of microcrystals of $CaCO_3$. Although kidney stones have been observed in rabbits, the $CaCO_3$ crystals do not usually lead to this problem. Urinary Ca concentrations are strongly dependent upon the dietary Ca content and, with a typical diet urinary excretion of Ca generally exceeds 2.5 mmol/day. This high urinary Ca excretion appears to be related to the high fractional intestinal absorption of dietary Ca since urinary Ca excretion declines to extremely low levels during dietary Ca deprivation. Consistent with the findings of high urinary Ca values, the fractional excretion of the Ca filtered by the kidney glomerulus in rabbits is much higher than the values of 2% or lower observed for most mammals. Nonetheless, as summarized in Table 1, renal Ca excretion is highly regulated, declining following PTH treatment and dietary Ca deprivation but rising following thyroparathyroidectomy and dietary phosphorus deprivation. PTH is believed to act through cAMP and PTH-sensitive adenylate cyclase activity has been localized to the proximal convoluted tubule and several of the more distal regions of the nephron (Chabardès *et al* 1975).

Table 1 Renal handling of calcium. Values are means ± S.E.M.

First author	Year	Fractional Ca excretion (%)	Influence of experimental manipulation
Berndt	1980	10±4	Increased to 25±7 following TPTX
Berndt	1980	23±7	Reduced to 7±1 with PTH treatment
Gilbert	1980	15±2	Reduced to 6±3 with PTH treatment
Bourdeau	1986	27±3	Reduced to 15±3 in vitamin D deficiency
Bourdeau	1988	37±2	Reduced to 4±1 with dietary Ca deprivation
DePalo	1988	36±5	Increased to 65±6 with dietary P deprivation
Pitts	1989	24±7	Reduced to 5±1 with PTH treatment

TPTX, thyroparathyroidectomy

Skeletal anatomy

Two aspects of rabbit bone structure are different from that of rodents. The growth plates of rabbit bones close by 32 weeks of age and osteonal (Haversian) remodeling of cortical bone occurs. Both of these features make rabbits a better model than rodents for studies of human bone metabolism. Osteonal remodeling can be visualized with radioisotopes such as ^{45}Ca and ^{90}Sr and with fluorochrome labels such as tetracycline, calcein and alizarin. Canaliculi can be visualized after silver impregnation of decalcified sections (Christie 1977).

Bone biochemistry and pharmacology

A number of proteins have been characterized in rabbit bone, including osteocalcin, osteopontin, receptors for insulin-like growth factor-I and vitronectin, a phosphotyrosyl phosphatase, a membrane-type metalloproteinase, and a cysteine proteinase. In agreement with its known ability to inhibit bone mineralization in other species, the bisphosphonate etidronate produces osteomalacia in rabbits. Chronic treatment with high doses of glucocorticoids inhibits bone formation and stimulates bone resorption.

Estrogen and bone

Rabbits are often characterized as a rapidly reproducing species and this fecundity results in part from their being induced ovulators. Female rabbits do not have menstrual or estrous cycles but ovulate following copulation. Thus, they do not have cyclical variations in circulating levels of estrogens and progestins that are eliminated by ovariectomy. Estradiol treatment of castrated male rabbits does not influence body length or weight but promotes growth plate closure while increasing vertebral bone mineral density by approximately one-third (Gilsanz et al 1988). This finding agrees with recent observations that estradiol has similar effects in men and thus has an important role in male skeletal development. The α form of the estrogen receptor has recently been identified in rabbit chondrocytes, osteoblasts and bone lining cells (Kusec et al 1998).

Reproduction

Gestation in rabbits lasts 32 days and lactation approximately 4 weeks. Mineralization of the fetal skeleton starts at day 16 (Danielson and Kihlström 1986) and at birth each kit contains about 7.5 mmol Ca. The rabbit placenta actively transports Ca as indicated by a fetal-to-maternal serum Ca gradient during late gestation. Lactation is unusual in that suckling occurs during a single daily episode lasting around 2 min. The Ca content of rabbit milk is very high, with values reported between 75 and 200 mM and these levels increase during the course of lactation (Lebas et al 1971). The retention of ingested milk Ca by the neonate is virtually complete. One study detected hypocalcemia in lactating rabbits fed a commercial diet, but alterations in serum Ca levels are not typically observed during reproduction. Serum calcitriol levels are elevated during lactation. Possible changes in the maternal skeleton during pregnancy and lactation have not been explored.

Osteopetrosis

A strain of spontaneously osteopetrotic rabbits was first identified in 1948 and the mutation responsible for this excess bone is known as 'os'. As is the case with other osteopetrotic mammals, the underlying defect is a failure of osteoclasts to resorb bone during skeletal development. Osteoclasts are present in bone of 'os-rabbits' but they are reduced in number and have an altered morphology. Since osteopetrosis cannot be cured by bone marrow transplantation from normal littermates, the defect causing this mutation appears to reside in the bone microenvironment rather than the osteoclast precursors (Popoff and Marks 1991).

Tumoral hypercalcemia

The VX2 tumor causes hypercalcemia when implanted subcutaneously in rabbits. Some evidence suggests that the hypercalcemia results from prostaglandin E_2-mediated bone resorption, but a more recent study indicated that excessive intestinal Ca absorption causes the hypercalcemia, as it was prevented by feeding rabbits a low Ca diet (Doppelt et al 1982).

References

Berndt TJ & Knox FG (1980) Effects of parathyroid hormone and calcitonin on electrolyte excretion in the rabbit. *Kidney International* **17** 473-478.

Bourdeau JE, Schwer-Dymerski DA, Stern PH & Langman CB (1986) Calcium and phosphorus metabolism in chronically vitamin D-deficient laboratory rabbits. *Mineral and Electrolyte Metabolism* **12** 176-185.

Bourdeau JE, Bouillon R, Zikos D & Langman CB (1988) Renal responses to calcium deprivation in young rabbits. *Mineral and Electrolyte Metabolism* **14** 150-157.

Butters RR, Chattopawdhyay N, Nielsen P, Smith CP, Mithal A, Kifor O et al. (1997) Cloning and characterization of a calcium-sensing receptor from the hypercalcemic New Zealand white rabbit reveals unaltered responsiveness to extracellular calcium. *Journal of Bone and Mineral Research* **12** 568-579.

Chabardès D, Imbert M, Clique A, Montégut M & Morel F (1975) PTH sensitive adenyl cyclase activity in different segments of the rabbit nephron. *Pflügers Archives* **354** 229-239.

Christie KN (1977) The demonstration of canaliculi in sections of decalcified bone by a silver impregnation technique. *Stain Technology* **52** 301-302.

Danielson M & I. Kihlström I (1986) Calcification of the rabbit fetal skeleton. *Growth* **50** 378-384.

DePalo D, Theisen AL, Langman CB, Bouillon R, & Bourdeau JE (1988) Renal responses to phosphorus deprivation in young rabbits. *Mineral and Electrolyte Metabolism* **14** 313-320.

Doppelt SH, Slovik DM, Neer RM, Nolan J, Zusman RM & Potts JT (1982) Gut-mediated hypercalcemia in rabbits bearing VX_2 carcinoma: New mechanism for tumor-induced hypercalcemia. *Proceedings of the National Academy of Sciences USA* **79** 640-644.

Gilbert PJ, Schlondorf D, Trizna W & Fine LG (1980) Renal effects of parathyroid hormone in the rabbit. *Mineral and Electrolyte Metabolism* **3** 291-301.

Gilsanz V, Roe TF, Gibbens DT, Schulz EE, Carlson ME, Gonsalez O et al. (1988) Effect of sex steroids on peak bone density of growing rabbits. *American Journal of Physiology* **255** E416-E421.

Gilsanz V, Roe TF, Antunes J, Carlson M, Duarte ML & Goodman WG (1991) Effect of dietary calcium on bone density in growing rabbits. *American Journal of Physiology* **260** E471-E476.

Haddad JG, Fraser DR & Lawson DEM (1981) Vitamin D plasma binding protein: Turnover and fate in the rabbit. *Journal of Clinical Investigation* **67** 1550-1560.

Kremer R, Bennett HPJ, Mitchell J & Goltzman D (1982) Characterization of the rabbit renal receptor for native parathyroid hormone employing a radioligand purified by reversed-phase liquid chromatography. *Journal of Biological Chemistry* **257** 14048-14054.

Kusec V, Virdi AS, Prince R & Triffitt JT (1998) Localization of estrogen receptor-α in human and rabbit skeletal tissue. *Journal of Clinical Endocrinology and Metabolism* **83** 2421-2428.

Lebas F, Besançon P & Abouyoub A (1971) Composition minérale du lait de lapine. Variations en fonction du stade de lactation. *Annales de Zootechnolgie* **20** 487-495.

Pitts TO, Puschett JB, Rose ME & Zull JE (1989) Effects of selective oxidation of 1-34 bovine parathyroid hormone on its renal actions in the rabbit. *Mineral and Electrolyte Metabolism* **15** 267-275.

Popoff SN & Marks SC (1991) Congenitally osteosclerotic (os/os) rabbits are not cured by bone marrow transplantation from normal littermates. *American Journal of Anatomy* **192** 274-280.

Tan SQ, Thomas D, Jao W, Bourdeau JE & Lau K (1987) Surgical thyroparathyroidectomy of the rabbit. *American Journal of Physiology* **252** F761-F767.

Warren HB, Lausen NCC, Segre GV, El-Hajj G & Brown EM (1989) Regulation of calciotropic hormones *in vivo* in the New Zealand white rabbit. *Endocrinology* **125** 2683-2690.

Wu DD, Boyd RD, Fix TJ & Burr DB (1990) Regional patterns of bone loss and altered bone remodeling in response to calcium deprivation in laboratory rabbits. *Calcified Tissue International* **47** 18-23.

Markers of bone resorption and formation during lactation in dairy cows

A Liesegang[1], M-L Sassi[2], J Risteli[2], M Kraenzlin[3], M Wanner[1] and J-L Riond[1]

[1]Institute of Animal Nutrition, University of Zurich, Winterthurerstr. 260, 8057 Zurich, Switzerland, [2]Department of Medical Biochemistry, University of Oulu, Finland and [3]Department of Endocrinology, University of Basle, Switzerland

Introduction

The goal of an optimal mineral supply is to achieve a well-balanced steady state between requirements and intake. At the initiation of lactation, calcium (Ca) homeostatic mechanisms are disturbed by a sudden increase in demand for Ca from 10 to 12 g to more than 30 g per day. Ca absorption from the gastrointestinal tract and Ca mobilization from bone need to be increased to re-establish homeostasis. In the second half of lactation the equilibrium is usually regained. The mobilization of Ca from bone can be described by using biochemical bone markers. Bone collagen degradation markers like hydroxyproline (HYP), pyridinoline (PYD), deoxypyridinoline (DPD), and cross-linked carboxyterminal telopeptide of type I collagen (ICTP) have been used in humans (Cosman *et al.* 1995, Eriksen *et al.* 1995, Risteli & Risteli 1997) and several animal species (monkey, horse, cow, sheep, pig, rat) for the monitoring of bone resorption, and osteocalcin (OC) has been used as a marker of bone formation (Scott *et al.* 1993, Egger *et al.* 1994, Cahoon *et al.* 1996, Scholz-Ahrens *et al.* 1996, Lepage *et al.* 1997, Liesegang *et al.* 1998).

The purpose of this study was to investigate if cows with a higher standard milk yield have a higher requirement for calcium originating from bone during lactation.

Animals, materials and methods

In a field trial involving 15 healthy brown Swiss cows with a mean standard milk yield of 6500 kg (group A) and 15 healthy brown Swiss cows with a mean standard milk yield of 4900 kg (group B), urine and blood samples were collected 14 days before parturition (a.p.), 14 days after parturition (p.p.), 1 month p.p., 1.5 months p.p. and monthly from 2 to 8 months p.p.

Calcium (Ca), magnesium (Mg), phosphorus (P), and urinary creatinine were determined by colorimetry with an autoanalyzer (COBAS MIRA® Roche-Autoanalyzer, Basel, Switzerland), using commercial kits. HYP concentrations in urine were measured with a colorimetric method (Bergman & Loxley 1963). DPD and PYD concentrations in urine were quantified with a HPLC method. OC and ICTP concentrations in serum were measured with a RIA. HYP, DPD, and PYD concentrations in urine were corrected for urinary creatinine. Changes with time of the concentrations in urine and serum were examined with the two-tailed Wilcoxon's sign rank test for paired comparisons and the group effect at the same time was investigated with the Mann-Whitney U test.

Results

No significant differences in mean serum Ca, Mg, and P concentrations were observed among the sampling days or between the groups during the 8-month collection period.

A significant increase of the corrected urinary HYP concentration was observed from 14 days a.p. to 14 days p.p. In the follow-up, HYP increased steadily until 3 months p.p. From 14 days a.p. to 14 days p.p., mean corrected DPD and PYD concentrations in urine increased significantly. Already 1 month p.p. concentrations of PYD and DPD began to decrease and reached the prepartum values of 1.5 months and 3 months p.p. respectively. No significant differences between the two groups were observed. Mean ICTP concentrations between the groups also did not differ significantly, but showed a significant increase 14 days p.p. (Fig. 1). One month p.p., ICTP concentrations returned to normal values. In contrast, mean OC concentrations decreased significantly from 14 days a.p. to 14 days p.p. and returned to prepartum values 1 month after parturition (Fig. 2). Mean 1,25-dihydroxyvitamin D concentrations peaked parallel to ICTP 14 days p.p. and returned to prepartum values already 1 month p.p. (data not shown).

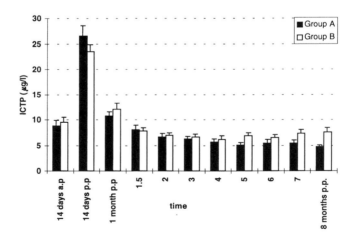

Fig. 1 Mean±S.E. ICTP concentrations during 8 months of lactation (group A: standard milk yield 6500 kg; $n=15$; group B: standard milk yield 4900 kg; $n=15$).

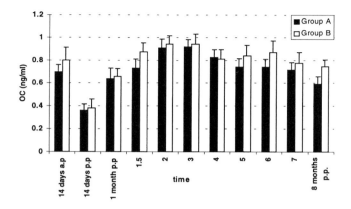

Fig. 2 Mean±S.E. OC concentrations during 8 months of lactation (group A: standard milk yield 6500 kg; $n=15$; group B: standard milk yield 4900 kg; $n=15$).

Discussion

The present study was undertaken to examine whether cows with a higher mean standard milk yield have to mobilize more calcium from bones than cows with a lower standard milk yield. The results do not demonstrate any differences between the two groups in relation to milk yield for all parameters analysed.

In conclusion, bone resorption markers indicate increased Ca requirements around parturition. This suggests that Ca requirements cannot only be covered through increased absorption from the intestines. It may be hypothesized that bone formation, which was measured with OC in this trial, is reduced to spare Ca at the beginning of lactation. Furthermore, 1,25-dihydroxyvitamin D is increased around parturition, which reflects increased Ca absorption from the intestines to cover the Ca demand at the beginning of lactation. Finally, these observations suggest that enough Ca was given in the diets during lactation and that the only critical period was around parturition.

References

Bergman I & Loxley R (1963) Two improved and simplified methods for the spectrophotometric determination of hydroxyproline. *Analytical Chemistry* **35** 1961-1963.

Cahoon S, Boden SD, Gould KG & Vailas AC (1996) Noninvasive markers of bone metabolism in the rhesus monkey: Normal effects of age and gender. *Journal of Medical Primatology* **25** 333-338.

Cosman F, Nieves J, Wilkinson C, Schnering D, Shen V & Lindsay R (1995) Bone density change and biochemical indices of skeletal turnover. *Calcified Tissue International* **58** 236-243.

Egger CD, Mühlbauer RC, Felix R, Delmas PD, Marks SC & Fleisch H (1994) Evaluation of urinary pyridinium crosslink excretion as a marker of bone resorption in the rat. *Journal of Bone and Mineral Research* **9** 1211-1219.

Eriksen EF, Brixen K & Charles P (1995) New markers of bone metabolism: clinical use in metabolic bone disease. *European Journal of Endocrinology* **132** 251-263.

Lepage OM, Eicher R, Uebelhart B & Tschudi P (1997) Influence of type and breed of horse on serum osteocalcin concentration, and evaluation of the applicability of a bovine radioimmunoassay and a human immunoradiometric assay. *American Journal of Veterinary Research* **58** 574-578.

Liesegang A, Sassi M-L, Risteli J, Eicher R, Wanner M & Riond J-L (1998) Comparison of bone resorption markers during hypocalcemia in dairy cows. *Journal of Dairy Science* **81** 2614-2622.

Risteli L & Risteli J (1997) Assays of type I procollagen domains and collagen fragments: problems to be solved and future trends. *Scandinavian Journal of Clinical and Laboratory Investigation* **57** (Suppl 227) 105-111.

Scholz-Ahrens KE, Barth CE, Drescher K, Goralczyk R, Rambeck WA, Wirner M *et al.* (1996) Modulation von Knochenmarkern in Plasma und Harn durch Nahrungscalcium beim ovariektomierten Göttinger Miniaturschwein. *Proceedings of the Society of Nutrition Physiology* **5** 77.

Scott D, Abu Damir H, Buchan W, Duncan A & Robins SP (1993) Factors affecting urinary pyridinoline and deoxypyridinoline excretion in the growing lamb. *Bone* **14** 807-811.

Cyclosporin A induces dentine resorption via a nitric oxide-cGMP pathway in osteoblast/osteoclast co-cultures: mechanistic insights

V S Shankar[1], S Yang[2], J Yang[2], M T Chapa[1], D J Simmons[2] and S Wimalawansa[1]

[1]Department of Internal Medicine, University of Texas Medical Branch, Galveston, Texas 77555-1065, USA and [2]Department of Orthopaedics and Rehabilitation, University of Texas Medical Branch, Galveston, Texas 77555-0892, USA

Introduction

Osteoporosis and its resulting fractures are often encountered in patients receiving cyclosporin A (CsA) for solid organ transplantations such as of the heart or liver. Intact and ovariectomized rats behave in a like manner (Movsovitz *et al.* 1988*a,b*). However, *in vitro* bone studies show that CsA inhibits bone resorption (Stewart *et al.* 1986, Klaushofer *et al.* 1987, Orcel *et al.* 1991). Our previous *in vivo* animal studies demonstrated that the effects of nitric oxide (NO) on bone are biphasic; at high doses, it increases bone resorption (S Wimalawansa, unpublished work). Therefore the variable effects of CsA may be due to evoked changes in the NO-cGMP pathway. In this study, we have extended these observations to an *in vitro* setting to determine whether the osteopenia caused by CsA administration is similarly tied to the NO-cGMP pathway.

Materials and methods

Osteoblast culture

Osteoblasts were isolated from the calvaria of 1-day-old Sprague-Dawley rats by six sequential 20 min digestions in Ca- and Mg-free collagenase solution. This procedure was repeated six times. The cells from the last three digests were pooled and cultured in minimal essential medium (αMEM) containing 10% fetal bovine serum and antibiotics.

Isolation and culture of osteoclasts

Newborn Wistar rats were killed by cervical dislocation; their femora and tibiae were removed; epiphyses were discarded. The diaphyses were opened with a scalpel, and the tissues were suspended in Hepes-buffered Medium 199 supplemented with fetal calf serum (10%, v/v). Osteoclasts were mechanically disaggregated by curetting the bones, agitating the suspension, and sedimenting cells on 22 mm 0-grade glass coverslips. Osteoclasts were preferentially adherent, and the other cells were washed away. The coverslips were placed in multiwell plates with Medium 199 containing 10% (v/v) fetal calf serum and incubated for 20 min at 37°C.

Preparation of dentine slices

Porpoise teeth obtained from the Galveston Stranded Mammal Program (Texas A&M University) were sectioned transversely at 100 mm; they were sonicated and sterilized. Four sections were used for each experimental modality.

Effect of NO donor on CsA-induced inhibition of osteoclastic dentine resorption

Freshly isolated osteoclasts were added to osteoblast-coated dentine slices and cultured for 48 h in a basal medium consisting of αMEM, 5% fetal bovine serum and 1% antibiotics (PenStrep, GIBCO, Grand Island, NY). CsA was applied to the co-cultures at concentrations of 1, 5 and 10 µg/ml; the NO donor L-arginine was applied at 1 mM. The cultures were then fixed in cold 95% ethanol and stained for tartrate-resistant acid phosphatase to identify osteoclasts and sites of osteoclastic resorption. A histomorphometric software package (Optimas Corp., Bothell, WA, USA) that projected a grid with points at 125 mm intervals (magnification ×65) was used to quantitate the percentage resorption.

Effect of NO synthase (NOS) inhibitor on NO donor reversal of CsA-induced inhibition of dentine resorption

The previous section was repeated with the addition of the NOS inhibitor N-ω-nitro-L-arginine methyl ester (L-NAME) (10 mM) before fixation with 95% ethanol.

Statistical analysis

Results are expressed as means±S.E.M. Comparisons between different treatment groups were made using ANOVA followed by Student's t-test. $P<0.05$ was considered statistically significant.

Results

The effects of CsA and the various treatment protocols on dentine resorption are summarized in Table 1. CsA attenuated resorption in a dose-dependent manner. The percentage of resorption at 1 µg/ml CsA did not differ from that seen in the controls. At 5 and 10 µg/ml, CsA significantly inhibited dentine resorption. Arginine by itself had little if any effect on dentine resorption. However, the CsA-induced inhibition of dentine resorption was antagonized by arginine. Increasing concentrations of CsA dose-dependently overcame the effects of arginine.

Blocking NO generation with L-NAME with or without arginine increased dentine resorption. Arginine once again sustained dentine resorption at a low concentration of CsA (1 µg/ml), but could not overcome the effects of CsA at higher concentrations.

Table 1 Effect of CsA on percentage detine resorption in osteoblast/ osteoclast co-cultures (n=4). The dose of arginine in each case was 3 mM and that of L-NAME was 0.1 mM.

Additions	Dose of CsA μg/ml	Mean	S.E.M.
Control	-	5.50	2.00
	1	4.97	1.8
	5	1.44*	0.22
	10	0.64*	0.09
Arginine	-	4.60	1.26
Arginine+CsA	1	12.5*	3.00
	5	6.60	1.40
	10	4.23*	1.50
Arginine+CsA+L-NAME	1	4.40	2.20
	5	1.10*	0.35
	10	0.55*	0.14

*$P<0.05$ vs control.

Discussion and conclusions

This study demonstrates that the bone-resorptive effect of CsA may seldom be evident *in vitro*. It has been proposed that the bone-resorptive system requires a replete endocrine system with reference to calciotropes, i.e. parathyroid hormine and vitamin D (Tullberg-Reinart & Hefti 1991). Nevertheless, we present two observations that pertain to the question of whether the CsA-mediated dentine-resorptive activity involves the NO-cGMP pathway. First, the concentrations of arginine (NOS substrate) normally present in the culture medium (0.9 mM) may have been able to inhibit low-concentration CsA-induced dentine resorption. Secondly, the NOS inhibitor L-NAME obviated the effects of arginine. Therefore it is possible that CsA inhibits osteoclastic resorptive activity, at least in part, via the NO pathway. The addition of the NOS inhibitor L-NAME negated the effects of the arginine, thereby exhibiting the same inhibition of resorption as demonstrated by the application of CsA only. These experiments are clearly preliminary to a test of the hypothesis that calcemic hormones are required to demonstrate the bone-resorptive effects of CsA *in vitro*.

References

Klaushofer K, Hoffmann O, Stewart PJ, Czerwenka E, Koller K, Peterlik M & Stern PH (1987) Cyclosporin A inhibits bone resorption in cultured neonatal mouse calvaria. *Journal of Pharmacology and Experimental Therapy* **243** 584-590.

Movsowitz C, Epstein S, Fallon M, Ismail F & Thomas S (1988a) Cyclosporin-A *in vivo* produces severe osteopenia in the rat: effect of dose and duration of administration. *Endocrinology* **123** 2571-2577.

Movsovitz C, Epstein S, Ismail F & Thomas S (1988b) Cyclosporin A in the oophorectomized rat: unexpected severe bone resorption. *Journal of Bone and Mineral Research* **4** 393-398.

Orcel P, Denne MA & de Vernejoul C (1991) Cyclosporin-A in vitro decreases bone resorption, osteoclast formation, and the fusion of cells of the monocyte-macrophage lineage. *Endocrinology* **128** 1638-1646.

Stewart PJ, Green OC & Stern PH (1986) Cyclosporin A inhibits calcemic hormone-induced bone resorption in vitro. *Journal of Bone and Mineral Research* **1** 285- 291.

Tullberg-Reinert H & Hefti AF (1991) Different inhibitory actions of cyclosporin A and cyclosporin A-acetate on lipopolysaccharide-, interleukin 1-, 1,25-dihydroxyvitamin D3-and parathyroid hormone-stimulated calcium and lysosomal enzyme release from mouse calvaria in vitro. *Agents and Actions* **32** 321-332.

Placental vitamin D regulates maternal calcium metabolism during pregnancy

H Fukuoka, M Haruna, C S Kim, M Fujita and Y Maekawa

Department of Developmental Medical Sciences, University of Tokyo, 7-3-1, Hongo, Bunkyo-ku, Tokyo, Japan

Introduction

In human, the calcium-regulating hormones, parathyroid hormone (PTH), 1,25-dihydroxyvitamin D_3 (1,25$(OH)_2D_3$) and calcitonin, strictly control serum calcium concentration. When serum calcium levels fall, PTH is promptly secreted by the parathyroid gland. It acts on renal tubules by stimulating the conversion of 25OHD_3 into 1,25$(OH)_2D_3$, the bioactive form of vitamin D. PTH and 1,25$(OH)_2D_3$ act together on peripheral organs to normalize serum calcium levels. Thus, in non-pregnant women hypercalcemic regulation is effected mainly by PTH and 1,25$(OH)_2D_3$, which is converted only in the renal tubules. In the serum, calcium is present mainly in association with albumin and in its free ionized form; the latter is the physiologically most important fraction, which is strictly regulated.

During pregnancy, the feto–placental compartment participates significantly in both maternal and fetal calcium metabolism (Hosking 1996) and thus pregnancy induces a very complicated shift in maternal calcium regulation (Boass et al. 1998). We studied calcium metabolism in pregnant women, and compared the results with data obtained from non-pregnant controls. We analyzed the changes in total and ionized calcium, assessed the physiological activity of the parathyroid gland, determined changes in serum 1,25$(OH)_2D_3$ and 24R,25$(OH)_2D_3$ concentration and the enzymatic conversion of vitamin D by the placenta.

Materials and methods

Subjects

Thirty-eight pregnant women, 22-28 years old, without any complications were selected. Five non-pregnant age-matched women served as controls. Informed consent to all subsequent studies was obtained from all subjects. Blood was venipunctured and serum was kept at −20°C until assay.

Placentas were categorized as either AFD (appropriate weight for date; delivery without any complications) or LBW (low birth weight; body weight ranging from 1958 to 2216 g). The delivered placentas were promptly dissected and the villous tissue was carefully scraped off with scissors on ice. The vitamin D conversion experiment is described below.

Serum total calcium and ionized calcium

Total calcium was analyzed by the ortho-cresolphthalein complexone (OCPC) method, and ionized calcium was measured with a calcium ion analyzer (Chiron634Ca^{2+}/pH Analyzer; Chiron, Tokyo, Japan).

Tubular resorption of phosphate (TRP)

The percentage TRP (%TRP) is a good biomarker for intact PTH activity. TRP for urine samples was calculated according to the following equation:

$$TRP = 1 - (UPO_4 \times serum\ CR)/(serum\ PO_4 \times UCR)$$

where UPO_4 is the urinary phosphate concentration, serum CR the serum creatinine content, serum PO_4 the serum phosphate concentration and UCR the urinary creatinine content.

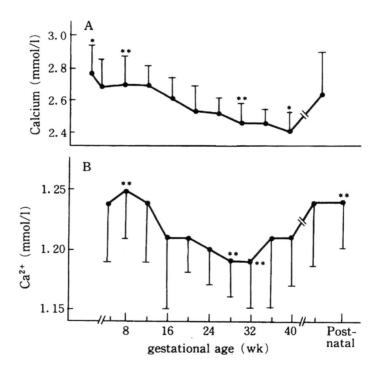

Fig. 1 Serum total (A) and ionized (B) calcium during pregnancy and puerperium. Results are mean±S.D. *$P<0.05$, **$P<0.001$ compared with normal.

Calcium Metabolism: Comparative Endocrinology

Vitamin D metabolites

$1,25(OH)_2D_3$ and $24R,25(OH)_2D_3$ levels in serum were assessed by RIA (Amersham, Tokyo, Japan).

Vitamin D conversion

[^{3}H]25(OH)D$_3$ was used as a substrate and incubated with the villous homogenate at 37°C for 60 min. After incubation, vitamin D metabolites were extracted, characterized and quantitated by reversed-phase HPLC analysis using pure $1,25(OH)_2D_3$ and $24R,25(OH)_2D_3$ as standards.

Results

Total and ionized calcium and TRP

During pregnancy serum total calcium declined until parturition (at term: 2.42±0.18 mmol/l). Serum Ca^{2+} levels first declined in parallel with total calcium, but after mid-pregnancy, started to rise through parturition and puerperium (Fig. 1). During gestation, values for %TRP ranged within the the limits for euparathyroidism, from 83 to 97% (Fig. 2).

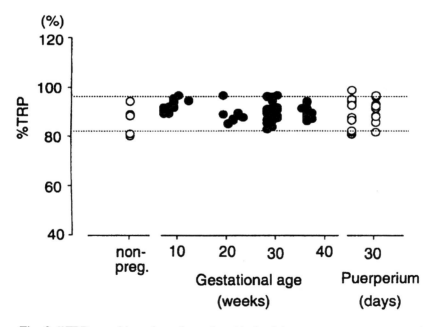

Fig. 2 %TRP as a biomarker of parathyroid gland in non-pregnant women and during pregnancy and puerperium.

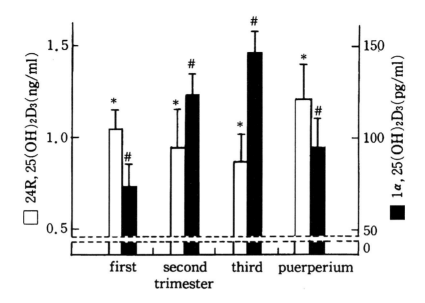

Fig. 3 Serum 1,25(OH)$_2$D$_3$ and 24R,25(OH)$_2$D$_3$ levels from first trimester until puerperium. *$P<0.01$ and #$P<0.01$ for 24R,25(OH)$_2$D$_3$ and 1,25(OH)$_2$D$_3$ respectively compared with values for non-pregnant women.

Vitamin D metabolites and conversion

Serum levels of 1,25(OH)$_2$D$_3$ increased until parturition and decreased again after delivery. Serum 24R,25(OH)$_2$D$_3$, however, showed a trend opposite to that of 1,25(OH)$_2$D$_3$ (Fig. 3). Lineweaver-Burk plot analysis of the vitamin D convertase activity in placenta (Fig. 4) yielded K_m values for 1α-hydroxylase and 24R-hydroxylase of 4.15 and 25.1 µM respectively; the V_{max} values were 0.9 and 4.3 ng/h per mg protein respectively.

Discussion

During pregnancy, when a physiological hydremia develops, serum albumin concentrations decrease until parturition (Boass et al. 1998), as did total calcium in this study. Interestingly, ionized calcium showed the same tendency to decrease but only until mid-pregnancy. After that, ionized and total calcium concentration dissociated, suggesting a strong upregulation. Surprisingly, our bioassay for PTH activity indicated that it remained constant during pregnancy and thus PTH cannot be responsible for the rise in ionized calcium after mid-pregnancy. PTH levels were within the normal range and thus pregnancy does not represent a state of hyperparathyroidism (Fig. 2), as has often been assumed (Frolich et al. 1991).

Calcium Metabolism: Comparative Endocrinology

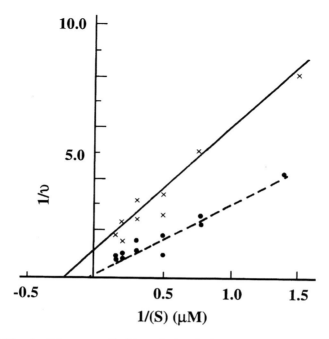

Fig. 4 Kinetic (Lineweaver-Burk) analysis of placental 1α-hydroxylase (×) and 24R-hydroxylase (●) activities. Velocity (v) is measured in ng/h per mg protein.

What then was the cause of the increase in maternal ionized calcium? We may obtain some clues from a report on two patients with gestational pseudohypoparathyroidism, features of which are primary derangement of renal 1α-hydroxylase and resistance of kidney and bone to PTH (Breslau & Zerwekh 1986). In these patients, PTH-receptor-mediated vitamin D conversion did not occur, resulting in a lowered concentration of $1,25(OH)_2D_3$ and serum calcium. During pregnancy, however, despite the primary derangement of the renal 1α-hydroxylase, serum $1,25(OH)_2D_3$ concentration did increase 2- to 3-fold and normocalcemia developed, while the serum PTH level was almost 50% of normal. After delivery, serum calcium and $1,25(OH)_2D_3$ decreased, and serum PTH rose. Apparently placental $1,25(OH)_2D_3$ contributed to the maintenance of maternal normocalcemia in these patients. As the physiological activity of the parathyroid gland does not change in normal pregnancy (Fig. 2), the renal PTH-mediated conversion of $1,25(OH)_2D_3$ may be similar to that of the non-gestational state. We showed the conversion of $1,25(OH)_2D_3$ and $24R,25(OH)_2D_3$ in the placenta. Compared with the 'non-pregnant' upper limit of 60 pg/ml, the concentration of $1,25(OH)_2D_3$ near term was extremely high. Placental $1,25(OH)_2D_3$ may have been released into the maternal circulation specifically for calcium homeostasis. The lowered calcium during gestation did not induce secondary hyperparathyroidism and this may result from suppression by elevated $1,25(OH)_2D_3$ levels of PTH production, through the vitamin D responsive element in the 5' promoter region of the PTH gene (van Leeuwen et al. 1992).

Our studies on the enzymatic activities of 25(OH)-1α-hydroxylase and 24R-hydroxylase in the normal term villi indicate the capacity of the placenta to produce significant amounts of $1,25(OH)_2D_3$ (0.80 ng/h per mg protein), which may explain the high maternal concentration of $1,25(OH)_2D_3$. Interestingly, we found a strong correlation between body weight of the newborn and the placental vitamin D convertase activities. We found that the placentas for LBW babies showed lower production of $1,25(OH)_2D_3$ and higher production of $24R,25(OH)_2D_3$ than the placentas for AFD babies, suggesting a role in regulation of fetal development. As $1,25(OH)_2D_3$ regulates active transfer of calcium across the cell membrane via a non-genomic action and induces the expression of 9 kDa and 28 kDa calcium-binding proteins in the placenta, the fetal development must be partly regulated by these enzymatic activities (Nikitenko et al. 1998). In addition, $1,25(OH)D_3$ stimulates basal cell proliferation in intestinal villi and increases the intestinal surface area to promote calcium absorption in the alimentary tract.

Thus, during pregnancy a large amount of $1,25(OH)_2D_3$ is produced by the placenta independently of maternal PTH and this metabolite regulates maternal calcium metabolism and possibly participates in fetal development. In other words, as for calcium metabolism during pregnancy, the placenta functions as the major organ.

References

Boass A, Garner SC, Schultz VL & Toverud SU (1998) Regulation of serum calcitriol by serum ionized calcium in rats during pregnancy and lactation. *Journal of Bone Mineral Research* **12** 909-914.

Breslau NA & Zerwekh JE (1986) Relationship of estrogen and pregnancy to calcium homeostasis in pseudohypoparathyroidism. *Journal of Clinical Endocrinology and Metabolism* **62** 45-51.

Frolich A, Rudnicki M, Fischer-Rasmussen W & Olofsson K (1991) Serum concentrations of intact parathyroid hormone during late human pregnancy: a longitudinal study. *European Journal of Obstetrics and Gynecological Reproductive Biology* **42** 85-87.

Hosking D (1996) Calcium homeostasis in pregnancy. *Clinical Endocrinology* **45** 1-6.

Nikitenko L, Morgan G, Kolesnikov SI & Wooding FB (1998) Immunocytochemical and In situ hybridization studies of the distribution of calbindin D9k in the bovine placenta throughout pregnancy. *Journnal of Histochemistry and Cytochemistry* **46** 679-688.

Van Leeuwen JP, Birkenhager JC, Vink-van Wijngaarden T, van den Bemd GJ & Pols HA (1992) Regulation of 1,25-dihydroxyvitamin D3 receptor gene expression by parathyroid hormone and cAMP-agonists. *Biochemical and Biophysical Research Communications* **185** 881-886.

A mid-region parathyroid hormone-related peptide (PTHrP) stimulates the absorption of calcium ions from the ovine rumen

D R Wadhwa[1], A D Care[1] and A F Stewart[2]

[1]Institute of Biological Sciences, Welsh Institute of Rural Studies, University of Wales, Aberystwyth SY23 3DD, UK and [2]Department of Medicine, University of Pittsburgh Medical Center, Pennsylvania 15213, USA

Introduction

Parathyroid hormone-related peptide (PTHrP) is initially translated as a preprohormone which is post-translationally processed to yield several secretory forms (Orloff *et al.* 1994), one of which is the N-terminal portion which bears a structural resemblance to the N-terminus of parathyroid hormone. Another secretory form is the mid-region portion (38-94), the structure of which is highly conserved. It has been demonstrated that PTHrP(38-94-NH_2) is active in four different biological systems, one of which is *in vivo* stimulation of placental transport of Ca^{2+} in the sheep (Wu *et al.* 1996). Moreover, this molecule circulates at a higher concentration in the ovine fetus (1 ng/ml) than in its mother (0.6 ng/ml; Care *et al.* 1997).

The reticulorumen has been shown to be an important site for the net absorption of Ca^{2+} in the adult sheep (Care *et al.* 1984, Holler *et al.* 1988). It is hypothesized that in the fetus it may feature in the reabsorption of Ca^{2+} after drinking of amniotic fluid containing Ca^{2+} of urinary origin. The present project was designed to test the efficacy of PTHrP(38-94-NH_2) as a stimulant of Ca^{2+} transport across adult ruminal epithelium and to locate the orientation of its putative receptor before carrying out a similar study using fetal ruminal epithelium.

Materials and methods

Pieces of ruminal epithelium from 12 sheep were mounted in eight computerized Ussing chambers, and the Ca^{2+} fluxes measured in each direction under short-circuit conditions (the potential difference between the two chambers clamped at zero) at 20 min intervals over a 3 h period using ^{45}Ca as tracer. Tissues were paired, on the basis of similar conductance, for the quantitative comparison of Ca^{2+} flux in each direction and the subsequent calculation of net flux. The conductances were monitored every 30 s to serve as a check for the maintenance of the integrity of the mounted membranes.

PTHrP(38-94-NH_2) was added to either the serosal or the mucosal surface of the epithelium after control values of Ca^{2+} flux had been established over approximately 1 h. After an equilibration period of 25 min, three or four measurements of Ca^{2+} flux in either direction were made to measure its effect on net Ca^{2+} transport. Other fragments of PTHrP tested at the serosal surface included PTHrP(67-86-NH_2), another proven stimulant of placental Ca^{2+} transport, and PTHrP(37-74), which has no known biological effects.

Concentrations of PTHrP(38-94-NH_2) of 20, 40 and 100 ng/ml (3, 6 and 15 nM respectively) were used in the serosal compartment but only a concentration of 100 ng/ml was used in the mucosal compartment. PTHrP(67-86-NH_2) was used at concentrations of 50 and 100 ng/ml in the serosal compartment, and PTHrP(37-74) was only used at a concentration of 100 ng/ml in the serosal compartment.

The mean unidirectional and net Ca^{2+} fluxes after the addition of a fragment of PTHrP were compared with the respective values obtained before the addition of each fragment. The significance of the results was tested using Student's paired t-test.

Results and discussion

Electrical properties of the rumen epithelium

The overall average conductance (G_t) and short-circuit current (I_{sc}) were 3.27 mS/cm^2 and 1.03 µEq/cm^2 per h, respectively. No significant changes in G_t and I_{sc} occurred during the experimental period, nor did the addition of a fragment of PTHrP change these electrophysiological parameters.

Effects of PTHrP(38-94-NH_2) on flux rates of Ca^{2+} ions across the epithelium

In all experiments, the Ca^{2+} flux in the mucosal to serosal direction during the control period was 2- to 3-fold higher than the Ca^{2+} flux in the opposite direction, resulting in a significant net Ca^{2+} influx. This flux represents active transport of Ca^{2+} in the mucosal to serosal direction. The addition to the serosal compartment of PTHrP(38-94-NH_2) at concentrations of 3, 6 or 15 nM (20, 40 or 100 ng/ml) significantly ($P<0.01$) increased the net Ca^{2+} flux, as a result of a significant increase in the mucosal to serosal flux; no significant change was noted in the magnitude of the flux in the opposite direction. The addition of 3 nM PTHrP(38-94-NH_2) increased the net Ca^{2+} flux by 42% whereas 15 nM PTHrP(38-94-NH_2) increased it by 56%. The onset of this stimulatory effect occurred within 20 min and it persisted for the duration of the experiment (usually 100 min). At the end of an experiment, the serosal and mucosal compartments were sampled for assay of PTHrP(38-94-NH_2) (Care et al. 1997). The serosal fluid concentration was found to be similar to that at the beginning. It was undetectable in the mucosal compartment, indicating that this peptide was not transported across the epithelium in a serosal to mucosal direction.

In contrast with the effect of the addition of 15 nM PTHrP(38-94-NH_2) to the serosal side, its addition to the mucosal compartment had no significant effect on Ca^{2+} fluxes in either direction. This suggests that the putative PTHrP(38-94-NH_2) receptor is located in the serosal surface of this epithelium.

The addition of either 100 ng/ml PTHrP(67-86-NH_2) or 100 ng/ml PTHrP(37-74) to the serosal compartment was without significant effect on the Ca^{2+} fluxes in either direction. This contrasts with the stimulatory effect of PTHrP(67-86-NH_2) on the placental transport of Ca^{2+} *in vivo* (Care *et al.* 1990). Perhaps the putative ruminal receptor for PTHrP(38-94-NH_2) is more discriminatory than its placental analogue.

Because the concentration (20 ng/ml) of PTHrP(38-94-NH_2) shown to stimulate active transport across ruminal epithelium from adult sheep is considerably higher than those circulating in either the fetus or the ewe, it is concluded at this stage that this peptide may be acting as a paracrine rather than an endocrine agent.

Acknowledgements

The award of a scholarship (to DRW) from the Association of Commonwealth Universities is gratefully acknowledged.

References

Care AD, Brown RC, Farrar AR & Pickard DW (1984) Magnesium absorption from the digestive tract of sheep. *Quarterly Journal of Experimental Physiology* **69** 577-587.

Care AD, Abbas SK, Pickard DW, Bari M, Drinkhill M, Findlay JBC *et al.* (1990) Stimulation of ovine placental transport of calcium and magnesium by mid-molecule fragments of human parathyroid hormone-related protein. *Experimental Physiology* **75** 605-608.

Care AD, Ratcliffe WA, Abbas SK & Stewart AF (1997) The role of calcitensin, parathyroid hormone related protein (38-94) amide, as a fetal hormone. *Journal of Endocrinology* **155** (Suppl. 2) 6.

Holler H, Breves G, Kocabatmaz M & Gerdes H (1988) Flux of calcium across the sheep rumen wall *in vivo* and *in vitro*. *Quarterly Journal of Experimental Physiology* **73** 609-618.

Orloff JJ, Reddy D, de Papp AE, Yang KH, Soifer NE & Stewart AF (1994) Parathyroid hormone-related protein as a prohormone: posttranslational processing and receptor interactions. *Endocrine Reviews* **15** 40-60.

Wu TL, Vasavada RC, Yang K, Massfelder T, Ganz M, Abbas SK *et al.* (1996) Structural and physiologic characterization of the mid-region secretory species of parathyroid hormone-related protein. *Journal of Biological Chemistry* **271** 24371-24381.

Part Five

Symposium Summary and Overview

Symposium summary and overview

H A Bern

Department of Integrative Biology, University of California, Berkeley
CA 94720-3140, USA

I am dubiously grateful to the organisers for the situation in which I now find myself. I am trying to act as a profound commentator in a field in which I have little competence. I am honoured that I should be invited despite this major defect. However, I am deeply appreciative of the daylong educational process: I found the programme excellently coherent and genuinely stimulating. Furthermore, I like the milieu and I like the people; what more can one ask for? You must all be exhausted by 10 hours of intellectual labour, plus the sopor-inducing food and wine, so that all you really need for immediate somnolence is an afterdinner talk. I shall aim at keeping it brief. Indeed, I considered presenting the briefest of overviews possible after Chris Dacke had given me a copy of the remarks of my predecessor, Professor David Fraser. Before I had seen them or even knew that they existed, I had tried to write down some items of possible central importance. Then I found most of them precisely stated by Fraser in his 1995 presentation - so much for so-called originality. I even pictured myself standing here, looking at your nodding heads and succinctly and clearly stating 'me too' and sitting down to your standing ovation. However, I have been scratching down comments as the meeting progressed to a point where I can barely decipher them.

The comparative approach is certainly where we are at. It is easy to recognise the ubiquity of the calcification process, as it is encountered in representatives of all living groups of organisms from bacteria to humans including protoctistans, fungi, plants and animals. So, the regulation of calcium, a vital component of all living matter, must involve homologous and/or analogous control systems, and the vertebrate/invertebrate line is really an artificial one in this particular context. I think Gert Flik's comments on crustaceans are germane to the whole field of calcium metabolism and its regulation. Comparative studies lead to generalities which are applicable on both the cellular/mechanistic and the organismal levels. If one looks back upon today's programme, we were indeed provided with a series of marvellous models. Throughout the Symposium the theme of evolution has shown its important face especially among the vertebrates. But I noticed a serious lacuna: what happened to the amphibians and the lungfishes? What happened to the water/land transition, which should play an important role in our thinking? Its absence from our considerations represents a real discontinuity.

Vertebrate calcium-regulating hormones are multiple, as if in their evolution organisms needed to provide insurance that calcium would be regulated reliably and precisely. Let me give you an example. Formerly it was possible to contrast fish with mammals, indeed Pat Ingleton used a diagram slightly familiar to me comparing humans with fish. The fish has stanniocalcin and calcitonin. The mammal has parathormone and calcitonin, and both have vitamin D derivatives. But this can no longer be taken at face value. Stanniocalcin is present in mammals, and calcitonin may not be an important calcium regulator in fishes, as Thrandur Björnsson indicated.

So, there is indeed multiple regulation, and comparative endocrinologists cannot afford to ignore potential regulators that used to be studied. I'll use Thrandur Björnsson's excellent discussion of the regulatory factors operative in fish to raise some questions. Björnsson listed various regulatory hormones in several categories. I would be less dismissive of that group of hormones that were placed under 'osmoregulatory factors'. Osmoregulatory hormones were considered *ipso facto* to be calcium regulators to some extent. Osmoregulation involves ion balance and that must include calcium. There is a host of osmoregulatory hormones. However, growth hormone, which is a proven osmoregulatory hormone in some teleost fishes, is also the hormone responsive for growth in a skeletal sense, and bone growth must involve calcium. Growth hormone itself is probably the major effective agent but acts through insulin-like growth factor which should be added to the listing. As for prolactin as an osmoregulatory hormone, one needs to re-examine its action in terms of a much more specific calcium-retaining role. It may be an effective agent in this area as we may have no reasonable substitute. However, we ought to look back upon the question of parathormone in the pituitary. This is an old issue which was not really resolved by the discovery of the third hypophysial member of the prolactin growth/hormone family: somatolactin. Somatolactin's role in calcium metabolism remains unclear. Early workers, such as Maurice Fontaine, raised the issue of a calcium-retaining factor in the pituitary at one of this kind of conference that occurred in the early seventies, and of course Peter Pang banged that drum for a long time. I don't know if anyone has ever proven that something like parathormone is really not present in fishes, and I mean parathormone not parathormone-related protein. The molecular tools are there to answer this question in regards to the pituitary, and I am not sure if that has been done. If parathormone-related protein in the fish pituitary is truly a secreted hormone, then we have a factor that may be playing the role once ascribed to a parathormone-like factor in the pituitary. Definitive answers to these questions are possible but they will not occur if we ignore the past of our field and dismiss these issues as if they were not still with us.

I agree with Dr Björnsson that osmoregulatory hormones as such are of significance in any overall consideration of calcium regulation. Other hormones act in a rapid fashion, a nongenomic action. If vitamin D derivatives act in part through nongenomic pathways, as was discussed at this meeting, we should not ignore other rapidly acting factors which play a role in water and ion balance in fishes, such as natriuretic hormones, angiotensins, neurohypophysial peptides, urotensins and so on, all of which act through nongenomic pathways and should not automatically be excluded as having effects on calcium metabolism. Cortisol is a prime osmoregulatory factor in fishes, but is also a stress hormone, and stress has effects upon calcium metabolism without doubt. I think this discussion leads to the recognition that there is a certain faddishness in regards to research in our field. One obviously wants to spend one's time with the most recent areas that look most exciting, where the technology is such as to allow one to go forward more rapidly than one has been able to in the past. However, I am not sure that comparative endocrinologists can afford an organismal approach which slights certain elements of the whole endocrine system.

Another aspect of comparative endocrinology that comes from fishes but applies to all vertebrates suggests that we need to pay more attention to the roles of hormones that are known to regulate calcium but which also have other important effects. If calcitonin is not a major calcium regulatory hormone in fishes, what about its other actions? The possible role of parathormone or at least its cleavage products in blood pressure regulation remains, and we did hear today about oviducal muscle regulation in birds and mammals. The pleomorphic aspects of vitamin D_3 indicate its possible action on processes other than calcium metabolism. At our meeting, estrogen and nitric oxide were also considered, but what happened to the thyroid gland? Hormone-binding proteins which help determine the activity of various agents add to the enormous complexity of the organismal picture of the endocrine regulation of calcium metabolism. It is difficult to generalize not only in regard to endocrine mechanisms but also in regard to physiological responses generally. There are so many sources of variation in regard to responses. Among determinants of physiological responses that were mentioned in this symposium are phyletic differences and species differences. Then there are individual differences, and one cannot overestimate the importance of the fact that even with a good group size, one or two animals simply do not do it the way the majority do it, and we are not sure why. Then there are sex differences, and age differences, and differences in reproductive state. In the case of fishes, there are salinity differences which occur among aquatic organisms generally; there is the possible intervention of stress and the hormones involved in stress, and of feeding and dietary factors, as emerged in our discussions today.

Calcium is a regulator of secretion, especially endocrine secretion. It may be specific for certain glands, but there is non-specific regulation of secretion not only of endocrine glands but also of non-endocrine glands. In its specific role, calcium is itself a chemical mediator and may always or most of the time act through a calcium-sensing receptor; in this case calcium has a parahormonal status. The calcium-sensing receptor is one of the few topics, I think, that Professor Fraser did not mention in his 1995 summary. We know at least in some organisms something about its gene or its protein structure. Pat Ingleton made a real contribution in pointing out what organs or what cells may have this receptor or what may not. There are still many unknowns, but the corpuscles of Stannius and the pituitary, as well as C cells and parathormone cells, may have this calcium-sensing receptor. What about the ultimobranchial cell? There are other aspects on the cellular level that are striking: I read a paper on the capacitative calcium entry system about which I knew absolutely nothing and do not know much more now. The molecular identity of this system was not reported. We need to learn about the hormonal regulation of the calcium-sensing receptor and of the capacitative calcium entry system. So I am impressed that good as you all are, you have not answered all of the questions! To these, I add another question. Many tissues, as well as organisms, show the ability to calcify, but some do not. I am not sure that this problem has been faced from the negative aspect; 'Why do some tissues not calcify? What is it that interferes with their expression of this common enough phenomenon?' I think that there are some answers that would come from such studies and that the pathophysiological implications are clear enough. The importance of continuing the efforts in which you have been involved is, I think, self evident. Studies of the kind that you have been doing have been dramatically productive to date and are truly promising for the future, not only in the provision of new data which by themselves do not represent new science but also in the provision of new concepts which do represent new science. So I wish you all great and continued success and thank you all for letting me be here.

Author index

Adebanjo O A 87
Anderson J 99

Balment R J 45
Ben-Bassat S 67
Bern H 185
Bi L X 145
Björnsson B Th 29
Brommage R 159
Bryden W L 103
Buford W L 145

Care A D 179
Carney D H 145
Chapa M T 169
Clement J G 49
Cook J 99
Coste H 139
Crowther R 145

Dacke C G 99
Danks J A 45, 49

Eberts M D 75
Elgar G 45
Estenik J F 113

Farquharson C 81
Faucheux C 131
Fischbach D G 75
Fisher S W 113
Flik G 3
Foster K 99
Fraser D R 93, 103
Fujita M 173
Fukuoka H 173

Gay C V 107

Haond C 3
Haruna M 173
Hubbard P A 45
Huff S N 75

Ikeya T 21
Ingleton P M 45, 49

Jahangir Z M G S 39
Jóhannsson S H 29
Johnson J V 87
Jüppner H 59

Kim C S 173
Kovacs C S 153
Kraenzlin M 165
Kronenberg H M 153
Kusuhara S 107
Kvetnansky R 39

Laing C J 93, 103
Lanske B 153
Larsson D 29
Lavelin I 67
Leach R M 67
Liesegang A 165
Lowe M 103
Lucu C 3

Maekawa Y 173
Malinski T 87
Manley N R 153
Markham G D 87
Martin T J 49
Matkovic V 75
Moonga B S 87
Morgan K M 75

Pârvu G 139

Author index

Patterson-Buckendahl P E 39
Persson P 21, 29
Pines M 67
Price J S 131

Riond J-L 139, 165
Risteli J 165
Rosol T J 75, 113, 119
Rubin D A 59
Rusnak M 39

Sandford R A 45
Sanz J 99
Sassi M-L 165
Shankar V S 169
Shea G M 93
Silverton S F 87
Simmons D J 145, 169
Stewart A F 179
Stromberg P C 113
Sugiyama T 107

Sundell K 29

Taylor J L 75, 113
Trivett M K 49
Turner R T 145

Waddington D 81
Wadhwa D R 179
Walker T I 49
Wanner M 139, 165
Watanabe T 21
Wheatly M G 13
Williams B 81
Wimalawansa S 169

Yang J 145, 169
Yang S 145, 169
Yarden N 67

Zaidi M 87
Zhang Z 13

Subject index

Page numbers in *italics* and **bold** indicate an illustration or table appearing away from its text.

A23187 13, 15
acidic matrix proteins, crustacean exoskeleton 22-4
Acipenser fulvescens 41
adrenaline (epinephrine) 39
aging, fracture healing in rats and 147, *148*
albumin, serum, in egg-laying turtles 115
alendronate 99
alkaline phosphatase (AP, ALP)
    chick growth-plate chondrocytes 69-70
    deer antler cells 133, 135, 136
alligator, vitamin D-binding protein 93-4
$\alpha\beta3$ integrin 107, 109
amiloride 7
aminoguanidine 87, 90
ammocoete (lamprey larva), PTHrP **51**, 53
amniotic fluid 156
amphibians 185
angiogenesis, in fracture healing 150
Anguilla rostrata see eel
antennal gland, crustacean 7, 16, 18
antlers, deer 131-7
    cultured cells 131
        alkaline phosphatase (ALP) activity 133, 135, 136
        PTHrP-stimulated [^{3}H]thymidine uptake 133, 135
        PTHrP synthesis 132-3, *134*
    PTH and PTH/PTHrP receptor expression 132, 133, *134*, 135-6
aragonite 22-3
Archosauria, vitamin D-binding protein 95, **96**
L-arginine 170, 171
arginine-glycine-aspartic acid (RGD), recognition by osteoclasts 107, 109, 110
asparaginase 159
Austropotamobius pallipes 5, **6**

birds
    Ca^{2+}-sensing and PTH receptor genes 67-72
    vitamin D-binding protein (DBP) 93-7
    see also chicken
bisphosphonates
    effect on avian medullary bone 99-102
    in rabbits 161
bladder
    air, Ca-sensing receptor 46
    urinary, Ca-sensing receptor 46, *47*
blastema, deer antler 131
    PTHrP and PTH/PTHrP receptor 133, *134*, 135
bluegill 40
bone
    anatomy, in rabbits 161
    biochemistry/pharmacology, in rabbits 161
    Ca balance, in mammals 122-3
    density, in PTH- and calcitonin-treated iguanas 78
    formation
        markers, in lactating cows 165-7
        in PTH- and calcitonin-treated iguanas 78, 79
    medullary *see* medullary bone
    morphometry, in PTH- and calcitonin-treated iguanas 76-7, 78, 79
    quality in chickens, dietary Ca and P and 82, 84, 85, 86
    resorption
        cyclosporin A actions *in vitro* 169-71
        dietary Ca and P content and 84-5
        in egg-laying hens 107, *108*-10
        in mammals 122-3
        markers, in lactating cows 165-7
        in PTH- and calcitonin-treated iguanas 78, 79
        regulation by nitric oxide (NO) 87
    teleost fish 32-3
bone sialoprotein 22
branchiostegites 9
bream, sea 40, 49

calbindin 30, 122
    placental 155
    in rabbits 160
calcein 75-6, 78
calcification
    absence in some tissues 188
    crustacean exoskeleton 21-2
        mechanisms 22-3
        mRNA expression during 23-4
calcified tissues
    teleost fish 32-3
    see also bone; scales
calcite 22-3
calcitonin (CT)
    in chickens, dietary Ca and P content and 84

Subject index

calcitonin (cont.)
  fetal 154
  fish 126-7, 186
    function 30, 187
  in invertebrates 10
  in mammals 119, 124-7
  placental Ca transport and 154
  in rabbits 160
  salmon 126-7
    administration in iguanas 75-80
calcitonin gene-related peptide 10
calcitriol *see* 1,25-dihydroxyvitamin D_3
calcium (Ca)
  balance, in mammals 120-1
  dietary content, broiler chickens 81-5
  excretion
    in mammals 120, 121-2
    in teleost fish 31-2
  fetal-placental metabolism 153-7
  haemolymph, in crustaceans 5, **6**
  milk, in rabbits 162
  parahormonal status 187-8
  serum/plasma
    in bisphosphonate-treated quail 101
    in egg-laying turtles 115
    fetal 127, 153, 157
    ionised *see* calcium ions (Ca^{2+}), serum/plasma
    in mammals 119-20
    measurement 120
    in phosphate-deficient pigs 140
    in pregnant women 173, 174, 175, 176
    in rabbits 159
    in tibial dyschondroplasia-susceptible birds 104-5
  urine, measurement 120
calcium absorption
  mammalian intestine 120-1, 122, 124
  ovine rumen 179-81
  rabbit intestine 159, 160
  teleost fish intestine 30-1
calcium-binding proteins (CaBPs)
  mammals 121
  teleost fish 29-30, 31-2
calcium (Ca^{2+}) channels
  crustacean gills 7
  fish 29, 30
  mammals 121
calcium ions (Ca^{2+})
  haemolymph
    in crustaceans 5, **6**
    hyperregulation in crustaceans 8, 9

  intracellular 120
  protein-bound, in mammals 119-20
  serum/plasma
    in bisphosphonate-treated quail 101
    dietary Ca and P content and 83, 84
    in mammals 119-20
    measurement 120
    in pregnant women 174, 175, 176, 177
    regulation in birds 67, 68, 72
    regulation in fish 4
    regulation in mammals 122, 123, 124
    set point 123
  uptake
    in crustacea 3, 5-10, 15
    in fish 4, 15
  see also calcium transport
calcium:phosphorus (Ca:P) ratio
  cortical bone, dietary influences in chickens 84, 85
  diet
    broiler chickens 81-5
    mammals 121
calcium pump (Ca^{2+}-ATPase) 13-19
  crustacean epithelia 7-8, 9, 13
  fish 13, **14**, 16
    gills 4, 30
    intestine 30-1
    kidneys 31
  future perspectives 18-19
  placental 155
  plasma membrane (PMCA) 13
    affinity (K_m) **14**, 15
    maximal flux rate (J_{max}) **14**, 15
    molecular characterization 16, **17**, 18
    physiological characterization 13-16
    temperature dependence 13, **14**, 15
  in rabbits 160
  sarco/endoplasmic reticulum (SERCA) 13
    molecular characterization 16-18
    physiological characterization 16
calcium-sensing cells, parathyroid gland 67
calcium-sensing receptor (CaSR, CaR) 187-8
  chicken 67
    gene expression 68, *69*
  fetal 153, 156, 157
  in fish 33, 45-8
    tissue distribution 46-8
  mammals 123
calcium transport
  crustacean gills 5-10
    apical membrane compartment 5-7

basolateral plasma membrane compartment 7-9
   regulation 9-10
  fish 29-34
   calcified tissues 32-3
   gills 4, 29-30
   intestine 30-1
   kidney 31-2
  mammals 121, 122
  placental 154-5
calmodulin, Ca^{2+}-ATPase regulation 13, 16, 18
calphotin 23
capacitative calcium entry system 187-8
carapace, crustacean 3
carp 30
cartilage
  deer antler
   PTHrP actions 135
   PTHrP and PTH/PTHrP receptor expression 133, *134*, 135-6
  fish, PTHrP **54**, 56
catfish (*Ictalurus punctatus*), PTH/PTHrP receptors 59
Chelonia, vitamin D-binding protein 95, **96**
chicken
  Ca^{2+}-sensing and PTH receptor genes 67-72
  dietary Ca and P content 81-5
  osteoclasts, nitric oxide production 87-91
  tibial dyschondroplasia *see* tibial dyschondroplasia
  vitamin D-binding protein 93
chief cells
  Ca^{2+}-sensing receptor (CaR) gene expression 68
  in calcitonin-treated iguanas 77
  in egg-laying turtles 115
  in mammals 123
  in PTH-treated iguanas *76*, 77, *78*
chloride cells
  Ca^{2+} uptake 29-30
  Ca-sensing receptor 46, *47*
chondrocytes
  chick growth-plate, PTH-R 69-71
  deer antler, PTHrP 135-7
  fish, PTHrP *55*, 56
cod, Atlantic (*Gadus morhua*) 30, 31
collagen
  degradation markers 165
  type I cross-linked carboxyterminal telopeptide (ICTP), in lactating cows 165, 166
  type II, chick growth-plate chondrocytes 69-70

type X, chick growth-plate chondrocytes 69-70
comparative endocrinology 185-8
corpuscles of Stannius 32
  Ca-sensing receptor 46-8
corticosterone 39
cortisol 187-8
cows, lactating, bone resorption/formation markers 165-7
crab
  Bermuda land 24
  blue (*Callinectes sapidus*) 5, **6**
  Chinese 7
  shore (*Carcinus maenas*) 5, **6**, 7-8
crayfish, freshwater 13-18, 22
cross-linked carboxyterminal telopeptide of type I collagen (ICTP), in lactating cows 165, 166
crustacea
  Ca^{2+}-ATPase 7-8, 9, 13-19
  Ca regulation 3-10
  exoskeleton 21-4
  gills, Ca^{2+} absorption 5-9, 15, 16
crypt cells *see* chloride cells
cyclic AMP (cAMP) 8
cyclosporin A (CsA), dentine resorption *in vitro* 169-71

DBP *see* vitamin D-binding protein
DD1 protein 23
DD4 protein 23
DD5 protein 23, 24
DD9 protein 23, 24
deer antlers *see* antlers, deer
denticles, dermal
  osteocalcin 40
  PTHrP **54**, *55*, 56
dentine, cyclosporin A-induced resorption 169-71
diet
  Ca:P ratio 81-5, 121
  Ca requirements, rabbits 159
  low phosphate, in pigs in Romania 139-42
  sources of Ca 120-1
differential display technique 23
1,25-dihydroxyvitamin D_3 (1,25$(OH)_2D_3$, calcitriol)
  in chickens, dietary Ca and P content and 83, 84-5
  dietary, tibial dyschondroplasia (TD) and 103
  in egg-laying turtles 115
  fetal 154
  free index, in tibial dyschondroplasia-susceptible birds 104-5

193

Subject index

in mammals 119, 124
placental Ca transport and 154
placental production 177-8
in rabbits 160
regulation of PTH synthesis 122
serum
   in lactating cows 166, 167
   in phosphate-deficient pigs 139-42
   in pregnant women 176, 177-8
   in tibial dyschondroplasia-susceptible birds 104-5
in teleost fish 31, 32-3
24,25-dihydroxyvitamin D_3 (24,25$(OH)_2D_3$)
serum, in pregnant women 176, 177-8
in teleost fish 31, 32-3
diltiazem 29
dogfish 49, 53

E2 *see* oestradiol-17β
ecdysone 18-19
ecdysteron 10
eel (*Anguilla rostrata*) 41
  Ca pump **14**, 15, 16
  Ca transport 30, 31-2
egg-laying cycle
  bisphosphonate administration and 99
  medullary bone formation 99
  parathyroid gland hyperplasia 113-15
  PTH and E2 effects on osteoclast adhesion 107-10
egg production
  in birds, Ca^{2+} metabolism 68
  role of vitamin D-binding protein 93
elasmobranchs
  osteocalcin 40-2
  PTHrP 49-56
endochondral ossification
  in deer antler regeneration 131, 136
  PTHrP and 56
endothelial cell stimulating angiogenic factor (ESAF) 145, 150
epinephrine (adrenaline) 39
epipodites 9
equilibrium potentials 5, **6**
Esox niger 41
estrogen *see* oestrogen
etidronate, in rabbits 161
exoskeleton, crustacean 21-4
  formation 21-2
  mechanisms of calcification 22-3
  mRNA expression during calcification 23-4
  organization 21

extracellular matrix proteins, acidic non-collagenous 22-4

female sexual maturation, teleost fish 32-3
fetus 153-7
  blood calcium 153
  calciotropic hormones 153-4
  calcium homeostasis 157
  kidneys 156
  parathyroid glands 156
  PTHrP 127, 154
  skeleton 156-7
fibroblast growth factor, basic (bFGF), fracture healing and 145, 146, 147-50
fish 29-34
  Ca^{2+}-ATPase *see* calcium pump (Ca^{2+}-ATPase), fish
  Ca^{2+} uptake 4, 15, 16
  calcium-regulating hormones 185-6
  calcium-sensing receptor 33, 45-8
  gills *see* gills, fish
  osteocalcin 39-42
  PTH 59
  PTH/PTHrP receptors 59-63
  PTHrP 33, 49-56, 59
flounder (*Platichthys flesus*) 45-8
foot restraint immobilization (IMMO) 39-40
fractures
  in broiler chickens, dietary influences 85
  healing in rats, effect of a thrombin peptide 145-50
  pathological, in captive reptiles 75, 79
  in PTH-treated iguanas 75, 79
frog 49
fruit fly, Ca pump **17**, 18

Gammarus pulex 5
geckos, vitamin D-binding protein 95, **96**
gene duplication 63
gestational pseudohypoparathyroidism 177
Gillichthys mirabilis 31
gills
  crustacean, Ca^{2+} transport 5-9, 15, 16
  fish
    Ca^{2+} transport/regulation 29-30
    Ca^{2+} uptake 4, 15, 16
    Ca-sensing receptor 46, *47*
    PTHrP **51**, *52*, 53
  phyllobranchiate 9
  trichobranchiate 9
glucocorticoids, in rabbits 161

granulomatous disease **126**
growth factors, fracture healing and 145
growth hormone 186
growth plates
   chondrocytes, PTH-R gene expression 69-71
   in rabbits 161

hagfish 49
hormones
   calcium-regulating 185-7
   osmoregulatory 186
   see also specific hormones
Hoxa3 gene knockout mice 155, 156
humoral hypercalcaemia of malignancy 119, **125**, 127
hydroxyapatite crystals 122-3
1α-hydroxylase, placental 176, *177*, 178
24R-hydroxylase, placental 176, *177*, 178
hydroxyproline (HYP), urinary, in lactating cows 165, 166
25-hydroxyvitamin D_3 (25(OH)D_3)
   conversion in pregnant women 175, 176
   DBP affinity in birds and reptiles 94-7
   dietary, tibial dyschondroplasia (TD) and 103
   free index, in birds and reptiles 94-7
hypercalcaemia
   familial hypocalciuric 123, **126**
   fetal 153
   idiopathic, in cats **126**
   of malignancy, humoral 119, **125**, 127
   tumoral, in rabbits 162
hyperparathyroidism
   neonatal severe 123, **126**
   nutritional secondary
      in captive reptiles 75, 78-9
      in mammals **125**
   primary **125-6**
   renal secondary **125**, **126**
hypervitaminosis A **125**
hypervitaminosis D **125**
hypoadrenocorticism **126**
hypocalcaemia, stress-associated **126**
hypodermis, crustacean 21
hypomagnesaemia **126**

iguanas (*Iguana iguana*) 75-80
   metabolic bone disease 75, 78-9
   PTH (1-34) or calcitonin administration 75-80
immobilization, foot restraint (IMMO) 39-40
Indian hedgehog 136-7
individual differences 187
insect cuticular proteins 24

inside out vesicles (IOVs), Ca^{2+}-ATPase 13-15
insulin-like growth factor 186
integrin, αvβ3 107, 109
intestine
   Ca absorption *see* calcium absorption
   fish
      Ca-sensing receptor 46, *47*
      Ca transport/regulation 30-1
invertebrates
   Ca^{2+}-ATPase 13-19
   Ca regulation 3-10

Jansen's chondrodysplasia 70
Japanese quail, medullary bone 99-102
jaw, fish, PTHrP **54**

kidneys
   fetal 156
   fish
      Ca excretion 31-2
      Ca-sensing receptor 46, *47*
      PTHrP **51**, *52*, 53
   mammals, Ca excretion 120, 121-2
   rabbits, Ca excretion 160-1

lactation
   in dairy cows, bone resorption/formation markers 165-7
   in phosphate-deficient pigs 140, 142
   in rabbits 162
lamprey (*Geotria australis*), PTHrP 49-56
Lepidosauria, vitamin D-binding protein 95, **96**
Lepomus macrochirus 39
lizards, vitamin D-binding protein 95, **96**
L-NAME (*N*-ω-nitro-L-arginine methyl ester) 90, 170, 171
lobster 7, 9
low birth weight (LBW) babies, placental 1,25(OH)$_2$D$_3$ production 178
lungfishes 185

mammals 119-27
   calcitonin 119, 124-7
   calcium 119-23
   calcium-regulating hormones 185-6
   1,25-dihydroxyvitamin D_3 119, 122, 124
   disorders of Ca metabolism **125-6**, 127
   PTH 119, 123-4, 137
   PTHrP 119, 127
   vitamin D-binding protein 93
medullary bone

Subject index

Japanese quail, effect of disodium pamidronate 99-102
  oestrogen-induced formation 99, 101
  osteoclasts
    isolation and purification 107-8
    PTH and oestrogen effects on adhesion 107-10
metabolic bone disease, in captive iguanas 75, 78-9
microsensor, nitric oxide measurement 87, 88-9
molluscs 4, 22-3
N^G-monomethyl-L-arginine 87
moulting cycle, crustacean 3, 18, 21-2
muscle cells, fish
  Ca-sensing receptor 46, 47, 48
  PTHrP 55

NADPH, chick osteoclast nitric production and 88, 89, 90
nerve cells, fish, Ca-sensing receptor 46, 47, 48
nifedipine 29
nitric oxide (NO)
  cGMP pathway, cyclosporin A actions 169-71
  porphyrinic microsensor measurements 87, 88-9
  production by chick osteoclasts 87-91
  regulation of bone resorption 87
nitric oxide synthase (NOS)
  inhibitors 90, 170, 171
  isoforms (eNOS, iNOS) 87
N-ω-nitro-L-arginine methyl ester (L-NAME) 90, 170, 171
norepinephrine (noradrenaline) 39
notochord, PTHrP 53-6

oestradiol-17β (E2)
  in male Japanese quail 99, 101
  medullary bone osteoclast adhesion and 107-10
  teleost fish 32, 33
oestrogen
  α receptor, in rabbits 162
  in rabbits 161-2
  regulation of PTHrP effects 137
  see also oestradiol-17β
Oreochromis mossambicus see tilapia
osmoregulatory hormones 186
osmoregulatory tissues, fish, PTHrP **51**, *52*, 53
osteoblast/osteoclast co-cultures, cyclosporin A-induced dentine resorption 169-71
osteoblasts
  rat, isolation and culture 169
  teleost fish 32

see also bone, formation
osteocalcin (OC) 22, 39-42
  in chickens, dietary Ca and P content and 83, 84-5
  circulating (pOC) 39
  effects of stress 39-40
  in fish 39-42
  serum
    in lactating cows 165, 166, *167*
    in phosphate-deficient pigs 139-42
osteoclasts 32
  calcitonin actions 124
  chicken medullary bone 107-10
    effects of PTH and E2 on adhesion 108-10
    isolation and purification 107-8
  isolated chicken, nitric oxide production 87-91
  rat, isolation and culture 169-70
  see also bone, resorption
osteogenic proteins (OP-1 and OP-2) 145
osteolysis, local **125**
osteomalacia **125**, 139
osteonal remodelling, in rabbits 161
osteopetrosis, in rabbits 162
osteopontin (OPN) 22
  chick growth-plate chondrocytes 69-70
  osteoclast adhesion 107
osteoporosis **125**, 169
oxygen radicals (O_2^-), measurement 91
oyster, pearl 22

Pacifastacus leniusculus **6**
pamidronate, disodium, effect on Japanese quail medullary bone 99-102
parathyroid gland (PTG)
  absence in fish 45
  adenoma or hyperplasia **125-6**
  chicken 67
    Ca^{2+}-sensing receptor (CaR) gene expression 68, 69
  in egg-laying turtles 113-15
  fetal 156, 157
  in PTH-treated iguanas 76, 77, 79-80
parathyroid hormone (PTH) 49, 186, 187
  bioassay, in pregnant women 174, 175, 176
  chicken
    Ca^{2+}-sensing receptor gene expression and 68
    in rickets 70
    in tibial dyschondroplasia 71
  in egg-laying turtles 115
  fetal 153, 154, 156

196

in fish 59
intact (PTH(1-84)) 123
in mammals 119, 123-4, 137
medullary bone osteoclast adhesion and 107-10
N-terminal (PTH(1-34)) 123
   administration in iguanas 75-80
placental Ca transport and 154
in rabbits 159
serum
   in phosphate-deficient pigs 139-42
   in pregnant women 177
parathyroid hormone (PTH)/PTH-related peptide receptors 67
cladistic relationships *61*, 62-3
in deer antler 132, 133, *134*, 136
in fish 59-63
human type-1 (hPTH1R) **62**, 63
human type-2 (hPTH2R) **62**, 63
knockout mice 136, 155, 156
in mammals 123
mutations 70
ovine rumen 180-1
in placental Ca transport 155
in rabbits 159
in rachitic chicks 69-71
in tibial dyschondroplasia (TD) 69-71
transgenic mice overexpressing 131, 136
zebrafish 59-63
   Southern blot analysis 62
   type-1 homologue (zPTH1R) 60-1, 62
   type-2 homologue (zPTH2R) 59-60, 62
   type-3 homologue (zPTH3R) 60, 61-2
parathyroid hormone-related protein (PTHrP)
in deer antler regeneration 131-7
fetal 127, 154
in fish 33, 49-56, 59, 186
knockout mice 70
in mammals 119, 127
mid-region (PTHrP(38-94)), Ca absorption by ovine rumen and 179-81
placental 155
placental Ca transport and 154-5
transgenic mice overexpressing 131
paresis
postparturient **126**
preparturient **126**
Penaeus japonicus (prawn), acidic matrix proteins 23-4
perichondrium, deer antler, PTHrP 133, *134*, 135-7
phosphate, inorganic (P_i)
dietary deficiency, pigs in Romania 139-42

milk, in phosphate-deficient pigs 140, 142
serum
   dietary Ca and P content and 83, 84
   in phosphate-deficient pigs 139-42
   in PTH regulation 123
tubular resorption (TRP), in pregnant women 174, 175
urinary, in phosphate-deficient pigs 140, 142
phospholamban 18
phosphophoryn 22
phosphorus (P)
dietary, in broiler chickens 81-5
serum, in egg-laying turtles 115
phyletic differences 187
pigs, low dietary phosphate 139-42
pituitary, fish
Ca-sensing receptor 46
PTHrP **51**, 53, 186
placenta
calcium transport 154-5, 162
   fetal regulation 154-5
   maternal regulation 154
   molecular mechanisms 155
vitamin D, regulation of maternal Ca metabolism 173-8
vitamin D convertase activity 176, 177-8
pore canals 21
prawn, kuruma, acidic matrix proteins 23-4
pregnancy, regulation of Ca metabolism 173-8
prolactin 186
protamine 159
protein kinase A 8, 18
protein kinase C 18
proton pump (H^+-ATPase) 7
PTH *see* parathyroid hormone
PTHrP *see* parathyroid hormone-related protein
puffer fish (*Fugu rubripes*) 45-8, 59
Puntius gonionatus 41
pyridinoline (PYD), urinary 165, 166

rabbits **17**, 159-62
5'-rapid amplification of cDNA ends (RACE) 60
rats
   Ca pump **14**, 15, **17**
   fracture healing, effect of thrombin peptide 145-50
rectal gland, elasmobranch, PTHrP **51**, *52*, 53
renal failure, acute or chronic **126**
reproduction, in rabbits 162
reptiles
   metabolic bone disease 75, 78-9
   parathyroid glands during egg-laying 113-15

Subject index

PTH and calcitonin administration 75-80
vitamin D-binding protein (DBP) 93-7
RGD sequences, recognition by osteoclasts 107, 109, 110
rickets
   in chicken
      Ca^{2+}-sensing receptor expression 68, *69*
      hypocalcaemic, dietary Ca:P ratio and *82*, 83, 84
      PTH-R gene expression 69-71
   in mammals **125, 126**
   in pigs in Romania 139
Romania, dietary phosphate deficiency in pigs 139-42
rumen, ovine 179-81
   electrical properties 180
   PTHrP(38-94-NH_2)-stimulated Ca absorption 179-81

saccus vasculosus, Ca-sensing receptor 46, *47*
salmon 30, 32
sarcolipin 16-18
scales, teleost fish 32-3, 40
SERCA *see* calcium pump (Ca^{2+}-ATPase), sarco/endoplasmic reticulum
sexual maturation, female teleost fish 32-3
sharks
   osteocalcin 40-2
   PTHrP 50-6
sheep rumen *see* rumen, ovine
skate (*Raja erinacea*), osteocalcin 40-2
skeletal tissues
   fish, PTHrP 53-6
   teleost fish, Ca balance/regulation 32-3
   see also bone
skeleton, fetal 156-7
skin
   deer antler (velvet), PTHrP and PTH/PTHrP receptors 133, *134*, 137
   fish
      Ca-sensing receptor 46, *47*
      PTHrP *55*
   human, PTHrP expression 137
skinks, vitamin D-binding protein 95, **96**
snakes, vitamin D-binding protein 95, **96**
sodium (Na^+), Ca^{2+}-ATPase and 13, 16
sodium/calcium ion (Na^+/Ca^{2+})-exchanger
   crustacean epithelia 7-8, 9
   fish gills 4, 30
   fish intestine 30-1
   regulation in invertebrates 9

sodium orthovanadate 13
sodium/proton ($2Na^+$/H^+)-exchanger, crustacean gills 6-7
sodium pump (Na^+,K^+-ATPase)
   crustacean epithelia 8-9
   regulation in invertebrates 9
somatolactin 33, 186
species differences 187
stanniocalcin (STC) 4, 186
   in animals other than fish 4
   Ca regulatory role 30, 33
stress 187-8
   -associated hypocalcaemia **126**
   effects on osteocalcin 39-40
swordfish 40

teeth, elasmobranch, PTHrP **54**, 56
teleost fish
   Ca balance and endocrine control 29-34
   osteocalcin 40-2
temperature sensitivity, Ca^{2+}-ATPase 13, **14**, 15
thiram 70-1
thrombin, synthetic peptide (TP508), fracture healing and 145-50
[^{3}H]thymidine, PTHrP-stimulated incorporation, in deer antler cells 133, 135
thyroid gland C-cells 124
tibial dyschondroplasia (TD)
   dietary Ca:P ratio and *82*, 83, 84
   genetic susceptibility 103-5
   PTH-R gene expression 69-71
tilapia **14**, 15, 16, 30
transbranchial potentials (TBP) 5, 6
transforming growth factor-β, fracture healing and 145
trout 30, 31, 32
   Ca pump **14**, 15, 16
   oestrogen-induced scale resorption 33
   PTH homologue 59
tumoral hypercalcaemia, in rabbits 162
turtles, snapping (*Chelydra serpentina*), parathyroid glands during egg-laying 113-15

ultraviolet-B (UV-B) light 75

verapamil 7, 29
vertebrae, fish, PTHrP **54**, *55*, 56
vertebrates
   Ca^{2+}-ATPase 13-19
   calcium-regulating hormones 185-6

see also birds; fish; mammals; reptiles
vitamin D 187
  deficient-chicks
    calcium-sensing receptor 68, *69*
    PTH-R 69-71
  metabolism, in rabbits 160
  metabolites
    in invertebrates 10
    in pregnant women 175, 176, 177-8
    in teleost fish 31, 32-3
    in tibial dyschondroplasia-susceptible chickens 103-5
    see also 1,25-dihydroxyvitamin D_3; 24,25-dihydroxyvitamin D_3; 25-hydroxyvitamin D_3
  placental, regulation of maternal Ca metabolism 173-8
vitamin D-binding protein (DBP)
  mammalian 93
  in rabbits 160
  reptiles and birds 93-7
    $25(OH)D_3$ binding affinity 94-7
    free 25(OH)D index 94-7
    plasma concentrations 94-7
  in tibial dyschondroplasia-susceptible chickens 103-5
vitamin D-binding protein/albumin/α-fetoprotein (DBP/ALB/AFP) common ancestor 93
vitamin D receptors
  in rabbits 160
  in tibial dyschondroplasia-susceptible chickens 105
vitellogenin 32
VX2 tumours, in rabbits 162

water/land transition 185

zebrafish, PTH/PTHrP receptors 59-63